Heating Ventilation and Air Conditioning

Fourth Edition

PRINTING . COPYRIGHT

Interface Publishing,
3 Station Lane, Greenfield,
Oldham, OL3 7EL, England.

First published 1999
Fourth edition September 2002

ISBN 0-9539409-3-4

Printed in Great Britain by Browns CTP, Oldham.
Tel. 0161 627 0010

REVIEWERS

Book reviews are an important tool for ensuring the validity of textbooks. This book has been reviewed by industry specialists. Their training, experience and knowledge of current trends place them in an ideal position for this task. I would therefore like to thank the following for agreeing to review, suggest changes and contribute, to sections of this book;

Roger Hitchin - BRECSU

Tony Bowen - Calorex Heat Pumps Ltd.

Marjorie Nicholson - Fenchurch Environmental Group

John Cooper - GasForce Ltd.

Andrew Clarke - Halton Products Ltd.

Chris Lincoln - Hamworthy Heating Ltd.

Doug Paterson - IMI Air Conditioning Ltd.

Rik Prowen - JS Humidifiers plc.

Reg Cross - Nationwide Filter Company Ltd.

John Ledger - Royair Ltd.

...and any other contributors who I have failed to mention. Any inaccuracies, poor English or other faults found in this publication are the responsibility of the author alone.

C O N T E N T S

H E A T I N G - S Y S T E M S

V E N T I L A T I O N

A I R - C O N D I T I O N I N G

INDUSTRY.PANELS

ENERGY.EFFICIENCY.ADVICE

ADDITIONAL.INFORMATION

Q U E S T I O N S

R E F E R E N C E

Heating.Ventilation.and.Air.Conditioning

INTRODUCTION

Building services is not a theoretical academic subject. It is a living, developing field of endeavour which touches everyone's lives. You are probably reading this in a heated (or air conditioned?) room right now! This book reflects the closeness between the academic study of building services with its practical application in its format. A traditional textbook is presented on the right hand pages and additional details including information on commercially available products and their suppliers is given on the left. By linking the theoretical descriptions with systems which can be seen around us in everyday life learning will be enhanced.

CONTRIBUTION FROM INDUSTRY

This publication represents a new way of supplying textbooks to students studying courses which have strong industrial links. It has been issued free of charge. All costs have been paid by members of the building services industry who, as professionals, are pleased to contribute to the education of the next generation of Architects and Building Services Engineers. The book can also be purchased by those outside the free circulation list at £19.99 from the publishers.

AIMS

- To give students access to a basic text at no cost to themselves.

- To introduce students to the basic concepts and components of heating, ventilation and air conditioning systems.

- To improve uptake and understanding of the subject by presenting photographs of commercially available equipment alongside the textbook description.

- To enhance the link between education and practice.

- To make students aware of the existence of companies and the range of products and services they offer at an early stage in their careers.

ACKNOWLEDGEMENTS

I would like to thank the many building services and environmental science lecturers throughout the country who have provided information on student numbers and have agreed to receive these books and distribute them to their students. As such a lecturer myself I know that the increased work load due to rising student numbers, reductions in funding, course development and research responsibility make any additional tasks difficult to accommodate.

I would like to thank all of the companies whose advertisements appear in this book and in particular the marketing managers and reviewers with whom I have liaised.

ABOUT THE AUTHOR

Richard Nicholls is a senior lecturer in the department of Architecture at Huddersfield University. He teaches environment and services on the Architecture degree and postgraduate diploma pathways and is pathway leader for the MSc. in Sustainable Architecture. He has experience of research as a research assistant investigating low energy housing in the department of Building Engineering, UMIST and industrial experience as a local authority Energy Manager. His most recent publication is a chapter on water conservation in the book "Sustainable Architecture" edited by Professor Brian Edwards

PUBLISHERS NOTE

The information given in this book is for guidance only. It is not intended to be exhaustive or definitive. All relevant standards, regulations and codes of practice should be consulted before any work is carried out.

1

Keeping tabs on Energy Efficiency

Energy efficiency advice

The main source of energy efficiency advice relating to buildings is the Carbon Trust's Action Energy programme.

The programme covers both the management and improvement of existing buildings and the design of new ones. It runs seminars and produces publications, including Case Studies and Energy Consumption Guides.

For building designers, the programme's most important activity is the Design Advice service. This provides professional, independent and subsidised advice on the energy-efficient and environmentally conscious design of buildings.

In this book seven information boxes dealing with energy efficiency issues have been sponsored by the Action Energy (see contents page for details). They will give you a flavour of the advice that is readily and freely available.

Action Energy publications are available to students, in response to specific requests, but Design Advice is intended for live projects.

Examples of some key design points promoted by the programme are as follows.

Lighting
- Make good use of daylight.
- Light to appropriate levels but not more.
- Use efficient electric lighting and provide good controls.

Cooling
- Minimise cooling loads by shading.
- Consider design strategies that avoid the need for mechanical cooling.
- If this is not possible, use low-energy cooling systems.

Heating
- Provide optimum building insulation.
- Specify efficient equipment.
- Provide effective controls.

Ventilation
- Consider natural ventilation or low-energy mechanical ventilation first.
- Keep duct air velocities low.
- Provide effective controls and choose efficient fans.

Further information
For further information on building-related energy efficiency, all it takes is one free phone call.
Take action.
Call Action Energy today.
0800 58 57 94
www.actionenergy.org.uk

ACTIONenergy

CARBON TRUST
Making business sense of climate change

1.0 Introduction

During the heating season, from early autumn to late spring, the weather becomes too cold for comfort. The building fabric protects us from the climatic extremes to an extent, but not enough to provide the comfort levels that modern society has grown to expect. Comfort can only be guaranteed using a space heating system. Most buildings in Britain require some form of space heating for the majority of the year. Heating not only gives thermal comfort to the building occupants but also ensures their health and, in a working environment, contributes to their productivity. Finally, heating protects the building fabric from deterioration by driving away moisture and preventing frost damage.

A requirement parallel to the need for space heating is the need for a method of providing hot water for washing and bathing. Unlike the seasonal requirement for space heating, hot water is required all year round.

The basic principle behind heating systems is very simple. Heat is released by burning fossil fuels or by passing an electric current through a wire. This heat is used to warm the occupants by radiant or convective means (see IP9). Whilst the principle is simple the functions must be carried out in a manner that ensures the following are satisfactorily considered;

- Economy - There are various costs associated with heating that must be minimised. These are, initial capital cost, maintenance costs and running costs. During a typical twenty year life of a heating system running costs will outweigh the initial capital costs many times over.

- Safety - Heating systems use combustible fuels, operate at high temperatures and release asphyxiant flue gases. These hazards must be managed so that they do not present a risk to the building or its occupants.

- Comfort - (see IP5) It is not possible to make all the occupants of a building satisfied with the internal temperatures at the same time. This is because personal preferences vary. However the system should aim to make the majority of the occupants comfortable. To achieve this the heating system should provide design temperatures and then control them within a narrow band of variation within occupancy hours.

- Environment - (see page 8) The combustion of fossil fuels releases gases which contribute to global warming and acid rain. To limit the damage, the amount of pollutant gasses released per unit of heating must be minimised. Type of fuel used, combustion characteristics, control and efficiency all contribute to minimising the volume of gas released.

Heating systems can be categorised into one of two main types these are; indirect heating systems and direct heating systems. The differences between the two systems will be outlined below. Section 1.1 considers indirect heating and section 3.0 considers direct heating.

Direct heating systems use individual stand alone heaters in each room where heating is required. The most common form of direct heating is the use of gas, coal or electric heaters in a domestic property. The capital and installation costs of any heating system are determined by size and complexity. For small systems direct heating has a low initial cost and can be easily expanded at a later date. Control of individual heaters is simple to achieve but group control, because of the physical separation, is more complex. Each heater must be provided with its own fuel supply and flue. Direct heating is extensively used as a cost effective form of heating in domestic, industrial and commercial buildings.

Indirect heating systems are known as central heating systems in houses because they generate heat at a central location, the boiler. The heat must then be removed from the boiler and delivered to each room. It is carried there by a heat transfer medium, which can be water, steam or air. Pipes are used to direct the flow of steam and water and ducts guide the movement of warm air. Heat emitters such as radiators (section 1.4) are required in the rooms to "hand over" the warmth from

IP1 - INSULATION.OF.DISTRIBUTION.PIPEWORK

Pipework is required to carry fluids as hot as 150°C (hthw) and as cold as -20°C through both heated and unheated spaces. The outcome of this is heat loss from hot pipes and heat gain by and condensation on cold pipes. Both conditions eventually result in a lack of system control and thermal discomfort.

HEAT LOSS FROM HOT PIPES

The heat loss rate from a pipe depends predominantly on its surface area (length of pipe run and pipe diameter), the temperature difference between it and its surroundings and the thermal conductivity of the pipe and any insulation materials surrounding it. Given that the pipe length, diameter and fluid flow temperatures are fixed by heating system design considerations, the element we can modify to reduce heat losses is the level of insulation around the pipe. This is recognised by the building regulations and water bylaws which lay down regulations governing the use of pipe insulation.

Increasing the thickness of the layer of insulation increases its resistance to the flow of heat. However, the cost also increases. The cost effective thickness of insulation must be determined from a knowledge of system design characteristics, fuel costs and insulation costs. In addition to insulating pipes it is necessary to insulate valves and other pipe fittings such as suspension rods. Specialist jackets are available for this purpose or sheet materials can be used by cutting and forming them into an appropriate shape.

Inadequately insulated pipework causes energy wastage, condensation risks, thermal discomfort and lack of system control

CHILLED WATER PIPEWORK

Heat gains by chilled water pipework must be considered in a similar manner to heat losses. However an additional feature which must be considered is the possibility of condensation forming on the cold pipework. To avoid this moisture laden air must be kept away from the cold surface of the pipe or any layer within the insulation which is at or below the dew point temperature. This is achieved using closed cell insulation products which have a high resistance to the passage of water vapour and by sealing any joins made in the material.

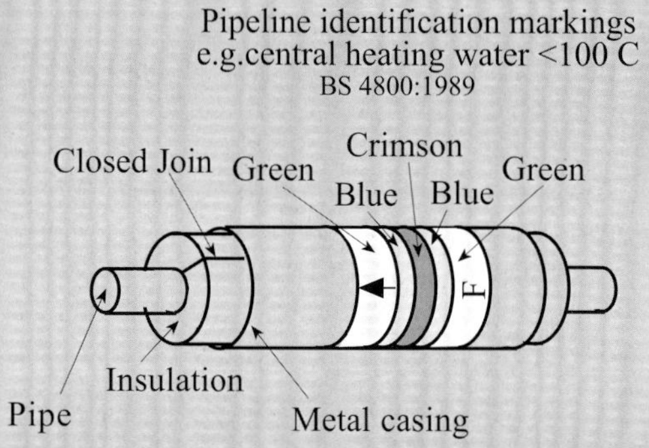

Figure IP1: Pipe insulation and identification

CONTROL AND COMFORT

Pipework is required to carry hot or cold fluids from plantroom to location of use such as a heating or cooling coil. Any heat lost or gained by the pipework will change the temperature of the fluid. As a result temperature sensor readings located early in the pipework may no longer reflect delivery temperatures. The absence of reliable sensor information makes control difficult. In addition to this thermal discomfort may be created as rooms overheat due to unregulated heat lost from pipes running through occupied spaces.

THE OZONE LAYER

chlorofluorocarbons (CFCs) (see IP15) are no longer used to make foamed pipe insulation as they damage the ozone layer. Environmentally responsible manufacturers now use ozone benign blowing agents such as air or carbon dioxide.

the heating system to the room air.

The advantages of indirect heating systems arise from the fact that most of the equipment is concentrated at a single location. This means only one flue and one fuel supply are needed to satisfy the entire building. This centrality means it is also possible to achieve a high level of control over the entire system

There are many types of building with various functions such as domestic, retail, industrial, educational and commercial. Within each of these categories there are different forms, fabric and heat loadings. Because of this it is impossible here to describe suitable heating systems to suit all buildings. This book will simplify matters by referring to two basic types of building: *domestic* which refers to housing and *commercial* which are buildings larger than domestic such as offices. In section 1.1 the components which make up an indirect heating system will be described using domestic central heating as a basis. Detail will also be given on how the domestic system is modified to satisfy the heating demands of large buildings.

1.1 Wet Indirect Heating

A wet indirect heating system uses water as the heat transfer medium. The main components of a wet indirect heating system are shown schematically in figure 1.1. Each of these components will be discussed more fully in sections following the numbering on the diagram.

In domestic properties it is likely that the boiler is accommodated in the kitchen either floor standing or wall mounted. In commercial buildings where the quantity of heating plant is greater it is necessary to have a purpose built plant room. The plant room must be well ventilated and have sufficient room for the equipment and access to it for maintenance. Plant rooms are usually situated on the ground floor where the weight of the heating system can easily be supported. Ventilation air for the plant room is typically provided through an access door leading to the outside which has louvered openings. Initial estimates of plant room size are based on rules of thumb, usually a small percentage of the total floor area. The actual percentage varies depending on the complexity of the heating system. As well as space for the plant room allowance must also be made for horizontal and vertical service distribution runs throughout the building [see IP7].

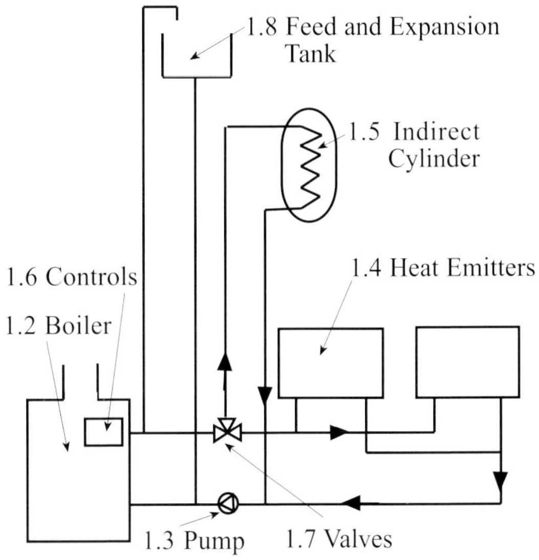

Figure 1.1 Wet indirect heating system

A number of companies produce packaged plant rooms which are delivered to site pre-assembled resulting in savings in both time and costs. These units may be accommodated within the building or are containerised for locating outside the building or on the roof top.

1.2 Gas Boiler

The heart of an indirect space and water heating system is the boiler. This section will concentrate on gas boilers however it should be remembered that oil fired boilers are available using liquid instead of gaseous fuels. The disadvantage of oil boilers is that oil deliveries must be organised and space allocated for oil storage. Electric boilers are also available, here the system water circulates over an electrical heating element. Electricity is more costly than gas or oil per unit of energy but the system has the advantage of small physical size, no requirement for a flue and ability to function where no gas supplies exist.

A gas boiler is a device which burns gas in a controlled manner to produce heat. This heat is transferred, using a heat exchanger, to water which circulates

around the heating circuit.

Boilers are specified in terms of their power measured in kilowatts (kW). A boiler for a typical four bedroom detached house would be rated at approximately 15 kW. A small flat or low energy house may need a boiler as small as 5 kW. Large buildings would need several hundred kilowatts of boiler power.

The main components of a boiler are shown in figure 1.2 and are described below.

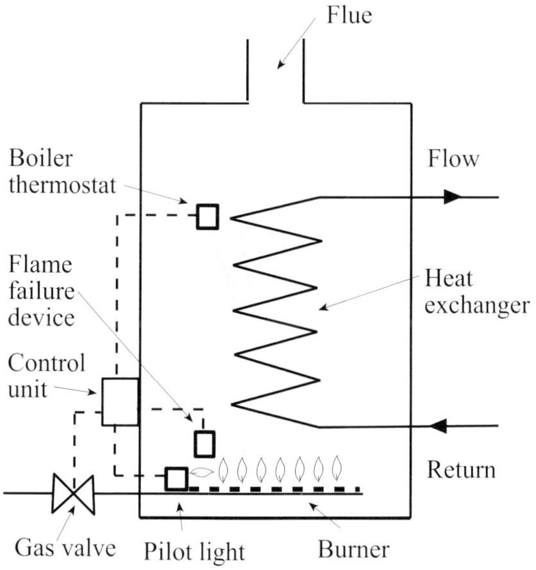

Figure 1.2 Parts of a boiler

Gas Valve. This valve is normally closed. It is opened by a solenoid allowing gas to flow into the burner if there is a call for heat and safety conditions are satisfied. If there is a loss of power or a signal indicating a fault is sent from the control unit then the valve will automatically close.

Pilot Light. This is a small flame which burns continually. Its function is to ignite the gas/air mix as it leaves the burner. An alternative is to use electronic spark ignition. Electronic ignitions have a greater degree of technical complexity but give improved boiler economy by eliminating the gas used by the pilot light at times when the boiler is not required to fire.

Burner. For optimum performance all of the gas that enters the boiler must be burned. To achieve this the burner must mix the gas with the correct quantity of air. This is known as a stoichiometric mix of gas and oxygen. Insufficient air would result in incomplete combustion with carbon monoxide being produced and dangerous unburnt gas building up. Too much air and the combustion gases will be diluted and cooled. The burner is designed to mix gas and air to give the most safe and efficient combustion possible. Atmospheric burners use the pressure of gas in the mains and buoyancy in the flue to draw combustion air into the burner. Forced draught burners use a fan to input combustion air. This allows a greater degree of control over the combustion process resulting in reduced boiler sizes.

In commercial buildings large amounts of fuel are used so it is important to buy it at the lowest price. To help in fuel purchasing boilers supplying large buildings can be fitted with dual fuel burners. These can use either gas or oil. It is possible to switch between fuels to use the fuel which at that time is the cheapest.

Amongst the combustion products of gas or oil are various nitrous oxides collectively known as NO_x. NO_x is a polluting gas and so European regulations exist to limit the amount of NO_x produced by burners for a given heat output. In response to this burner manufacturers have updated their products to meet and often exceed the requirements of this legislation.

Flame Failure Device (FFD). Flames are detected by this unit and if present gas is allowed to enter the burner. If the flames are extinguished for any reason the gas valve will be closed. This avoids a dangerous build up of unburnt gas within the boiler.

Control Unit. This is an electronic device which receives signals from the FFD and Thermostats. Using this information it controls the operation of the gas valve, pump and ignition systems. Time control is also carried out by this unit to make sure that the heating system only operates when it is required. Each type of boiler has its own control strategy, individual boiler manufacturers should be consulted for further details.

Boiler Thermostat. This is a temperature sensor which is used to control the boiler flow temperature. Boiler

2

ACTIONenergy

Keeping tabs on Energy Efficiency

Environmental effects of energy consumption

The combustion of fossil fuels creates a number of pollutant gases. The quantity emitted depends on the amount and type of fuel being burned. The gases are carbon dioxide (CO_2), one of the main contributors to global warming; sulphur dioxide (SO_2), which contributes to the problem of acid rain; and nitrous oxides (NO_x). NO_x is a collective term for the oxides of nitrogen, mostly nitric oxide (NO) (a cause of ozone depletion) and nitrogen dioxide (NO_2). NO_2 is a lung irritant which reduces air quality.

Fuel Type	Kg CO_2 /kWh
Electricity	0.44
Oil	0.26
Natural Gas	0.19

Carbon Dioxide Emission Factors

Global warming

Heat from the sun can easily pass through the atmosphere to warm the Earth's surface. However, the heat re-radiated from this surface is trapped by gases in the atmosphere, such as CO_2. This causes the atmosphere to be warmed (global warming) .Some global warming is essential for life on Earth.

However, recent decades have seen an increase in the concentration of CO_2 in the atmosphere as a result of fossil fuel burning. Scientists now generally agree that increased heating of the atmosphere will occur due to the effects of global warming, leading to climate change.

Acid rain

Some of the particulates and gases produced during combustion are acidic and combine with rain to create acid rain. This damages the environment in three ways.

- It harms trees – many European forests are showing signs of leaf damage due to acid rain.
- It collects in freshwater lakes and rivers, increasing the acidity which endangers freshwater life.
- It erodes stone buildings and statues, degrading the beauty of historic buildings.

The role of the building professionals

The first stage in reducing the problem is to minimise the consumption of fossil fuels while maintaining acceptable standards of comfort. Energy use in buildings accounts for approximately 45% of the UK CO_2 output.

Further information

For further information on building-related energy efficiency, all it takes is one free phone call.
Take action.
Call Action Energy today.
0800 58 57 94
www.actionenergy.org.uk

CARBON TRUST
Making business sense of climate change

flow temperature is the temperature of water leaving the boiler. It is this which determines the radiator temperature. The hotter the water the greater is the heat output of the radiators. It is usual to set the boiler thermostat higher in winter than summer because of this. This process is carried out manually on domestic boilers however commercial boilers are fitted with a device called a compensator which carries out the function automatically. Compensators are discussed more fully on page 35. A separate overheat thermostat provides a safety function by cutting out the burner if the temperature should increase too much. Thus avoiding overheating of the boiler.

Heat Exchanger. Made of materials such as cast iron, steel and aluminium the heat exchanger is designed to give maximum thermal contact between the hot combustion gases and the circulating water. The heat exchanger of large boilers may be delivered to site in sections which are then bolted together. There is a variety of heat exchanger forms. Some are positioned over the burner and the hot flue gases rise up through the heat exchanger. Other heat exchangers surround the burner and the combustion gases have to pass through channels in the heat exchanger to escape. This means the flue gases have to pass the heat exchanger twice thereby improving the transfer of heat into the heating circuit.

Flue. When gas is burnt in air carbon dioxide, carbon monoxide, nitrous oxides and water vapour are produced. Carbon monoxide is an asphyxiant and would kill the occupants of any room in which it accumulated. To avoid this, it is necessary to have a flue which carries away the waste products of combustion and safely discharges them outside of the building. A flue is essentially a duct connected to the boiler combustion chamber, terminating outside of the building. The flue run should be as straight as possible to avoid unnecessary restriction to the flow of flue gas. Horizontal runs should be avoided to allow flue gases to rise continuously. It should terminate at a location where the flue gases cannot re enter the building. Hence, for example, flues cannot discharge near windows. The flue will also have a terminal unit which acts to keep the outlet of the flue open by, for example, excluding the entry of nesting birds. Flue gases are hot and so the terminal should not be located in a position where it could be touched by anyone passing by. The movement of flue gases in atmospheric burners is by natural buoyancy. Forced draught boilers use fans to discharge the products of combustion

There are various arrangements of flue. However, each one exhibits the common functions of safely exhausting flue gases whilst at the same time preventing the burner flames being blown out by excessive draughts through the system. One method of preventing this is to use a draught diverter (figure 1.3). Upwardly moving flue gases pass up and around a baffle plate. If wind causes the direction of flow to reverse then the plate causes the flue gases to temporarily spill over into the boiler house rather than enter the boiler. The draught diverter also prevents excess air being pulled through the boiler should there be excess suction from the flue itself due to wind or buoyancy effects.

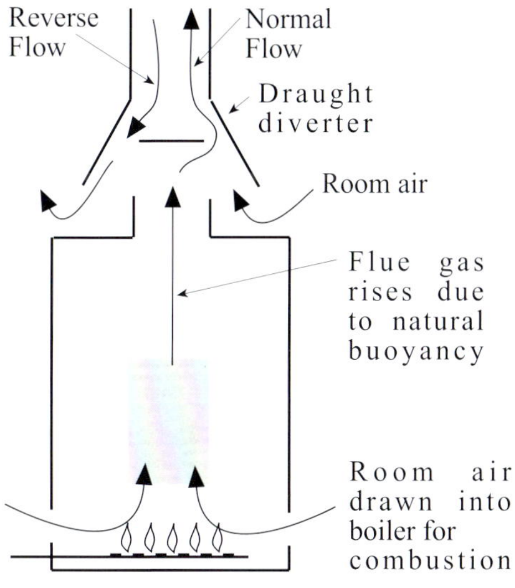

Figure 1.3 Flue draught diverter

The Balanced Flue (Figure 1.4) is a dual function flue which takes combustion air from outside the building and supplies it to the burner. The same unit discharges the flue gases outside. The benefit of this flue is that any wind pressures act on the inlet and outlet equally. As a result flows through the flue will be stable ensuring the burner flames will not be blown out. Because both the combustion air and flue gases enter and leave the boiler without making contact with the room air the boiler is known as a "room sealed" appliance.

I P 2 - TEMPERATURE, ENERGY. AND. POWER

Three terms commonly used in building services studies are temperature, energy and power. The latter two can easily be confused. This information panel aims to clarify the definition of these terms and give examples related to buildings.

TEMPERATURE

The scale of temperature commonly used in building studies is the Centigrade scale. This is a scale set between the temperature of melting ice and the temperature of boiling water. These temperatures are zero degrees centigrade (0°C) and one hundred degrees centigrade (100°C) respectively. Typical temperatures encountered in buildings are;

Design outside air temperature, -1°C
Average annual outside air temperature, 6°C
Chilled water flow temperature, 6°C
Room temperature (active e.g. gymnasium), 16°C
Room temperature (sedentiary e.g. office), 21°C
Human core body temperature, 37.5°C
Boiler flow temperature, 82°C
Boiler return temperature, 70°C
Max. temperature of radiant tube heater, 450°C

Another scale of temperature used by building scientists, and one which you may encounter, is the Kelvin scale of temperature. The divisions on this scale are exactly the same as on the centigrade scale i.e a change (Δ) of one degree centigrade is equivalent to a change of one degree Kelvin ($\Delta 1^{\circ}C \equiv \Delta 1K$). The kelvin scale starts at 0K which equals -273°C so 0°C would therefore be equivalent to 273K.

ENERGY

Energy is thought of as the ability to do work. There are various forms of energy, for illustration they can be described in relation to a CHP unit (page 17). The forms of energy are; *chemical energy* as is contained in fuels such as coal, oil or gas, *mechanical energy* which is held by rotating objects such as the flywheel of a CHP unit, *thermal energy* (Heat) that is released by burning fuels and *electrical energy* which is produced by the CHP unit generator. Note that thermal energy is often simply referred to as heat

The amount of energy held in any of the above forms can be quantified. To do this we need units of energy. The basic scientific unit of energy is the Joule (J). But this unit is too small for describing the quantities of energy used in buildings. Instead we normally use the unit, Watt hour. This is still small so we use thousands(kilo (k)) of Watt hours i.e. kilowatt hour (kWh). 1kWh is equivalent to 36,000,000 Joules! or 0.036 Gigajoules (GJ). Typical energy values encountered in buildings are;

Energy used by a 1 bar electric fire each hour = 1 kWh
Energy used to heat a house for one year, 40,000kWh
Energy contained in 1m^3 of gas = 10.5 kWh
Energy contained in 1kg of coal = 9.02 kWh
Energy contained in 1 litre of oil = 10.4 kWh

Note when heat is added to an object its temperature increases. When heat is removed from a body its temperature decreases.

POWER

Energy cannot be created or destroyed but it can change from one form to another. The rate at which this change occurs in a system is called the *power* of the system. For example, a gas boiler is a machine to convert chemical energy (gas) to thermal energy (heat). This conversion is not instantaneous, it occurs over time. If, in a given time, a boiler converts more gas to heat than a second boiler, then the first boiler has a greater power.

The unit of power is the Watt (W) (1W=1Joule/second). Once again this is a small unit so we often use kilowatts (kW). Typical power values encountered in buildings are;

Light bulb =	100W
1 bar electric fire =	1 kW
Boiler for a low energy house =	4kW
Boiler for a detached house =	12 - 18 kW
Commercial boiler =	150kW
Domestic refrigerator =	150W

Typical split A/C unit = 0.8 to 3kW (electrical input)
giving 2.4 to 9kW of cooling

Ventilation and combustion air is required in rooms where non room sealed combustion appliances are operating. It is needed to supply sufficient air to allow complete and safe combustion of the gas. In large installations ventilation also helps to disperse unwanted heat build up. In housing, purpose provided ventilation for small (less than 7 kW) and room sealed appliances is not required. However for non room sealed and larger appliances purpose built air vents should be provided connecting the room to the outside air. There are exceptions and reference should always be made to current regulations and manufacturers data.

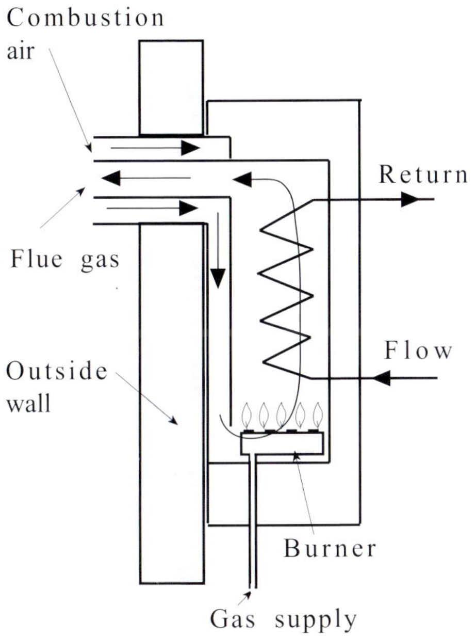

Figure 1.4 Balanced flue

In commercial buildings ventilation is usually provided through ventilation openings in the plant room walls or door. It can also be supplied to internal plant rooms using a fan and ducting running from outside to the plant room. Sensors in the ducting are interlocked with the boiler controls. These interlocks switch off the boilers if the ventilation air supply is stopped for any reason such as failure of the ventilation fan.

Fan Dilution is a flue system which cools and dilutes the flue gases so that they may be discharged at low level. The system works by drawing air from outside the building along a horizontal duct (figure 1.5). The boiler discharges its combustion gases into this airflow and so they become cooled and diluted. It is then possible to discharge the flue gases into a well ventilated area such as above the rear exit of a building. Dilution air inlet and flue gas outlet should preferably be on the same side of the building to avoid draughts blowing through the system.

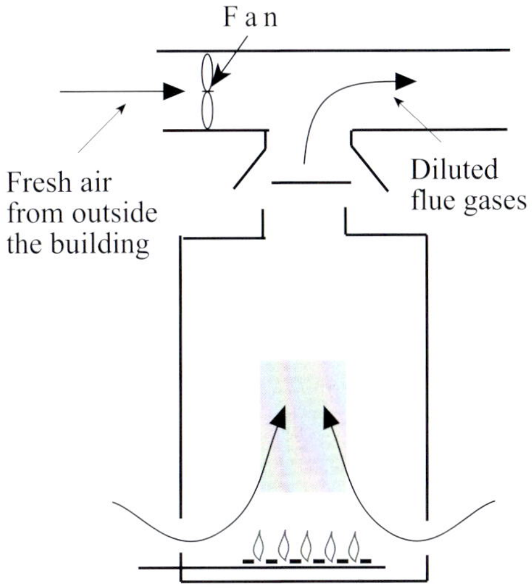

Figure 1.5 Fan dilution system

1.2.1 Boiler Efficiency

The efficiency of a boiler is a measure of how well it converts fuel to heat.

Boiler heat input is in the form of gas or oil. When this is burnt the aim is to transfer all of the heat that is released into the heating circuit. A system that achieved this aim would be 100% efficient. For safety reasons waste combustion gases must be cleared from the boiler. This is carried out by allowing the natural buoyancy of the hot flue gases to carry them up and out of the flue. Unfortunately the heat contained in these gases is lost to the system. As a result any flued combustion appliance can never operate at 100% efficiency. When selecting a boiler reference is made to manufacturers information contained in product data sheets. Figures for efficiency are usually given but if not it can be easily worked out from quoted heat input and output values (figure 1.6).

NOTES

Three classes of boiler efficiency can be identified;

Standard Boilers. A standard boiler is one which provides good quality utilitarian heating but has no cost increasing features that would enhance its efficiency. The efficiency of all boilers varies with the amount of work they are required to do, known as the boiler load. For this reason the average efficiency of a standard boiler over the heating season is usually given as the seasonal efficiency. For a standard boiler this is typically 75%. The variation of efficiency with load is discussed more fully in a later section.

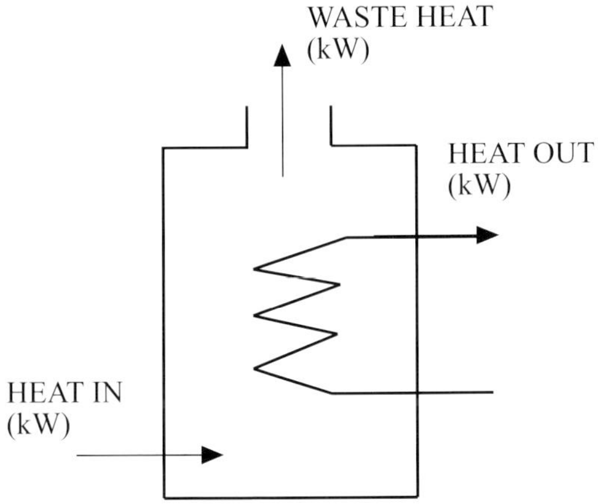

$$\text{EFFICIENCY}(\%) = \frac{\text{HEAT OUT}}{\text{HEAT IN}} \times 100$$

Figure 1.6 relationship between efficiency and heat output and input.

High Efficiency Boilers. These boilers are more costly than standard boilers because they include features such as a larger heat exchanger, additional casing insulation, electronic ignition and flue dampers (fig 1.7). These features; absorb more heat from the flue gases, reduce casing heat losses, stop gas usage when there is no call for heat and prevent convective loss of heat when the boiler finishes firing respectively. As a result the seasonal efficiency is improved to approximately 85%.

Condensing Boilers. These boilers have a high operating efficiency. This is due to their large heat exchanger which extracts so much heat out of the flue gases that the vapour in them condenses onto the heat exchanger (hence the name). In this way the heat exchanger recovers both sensible and latent heat from the flue gas.

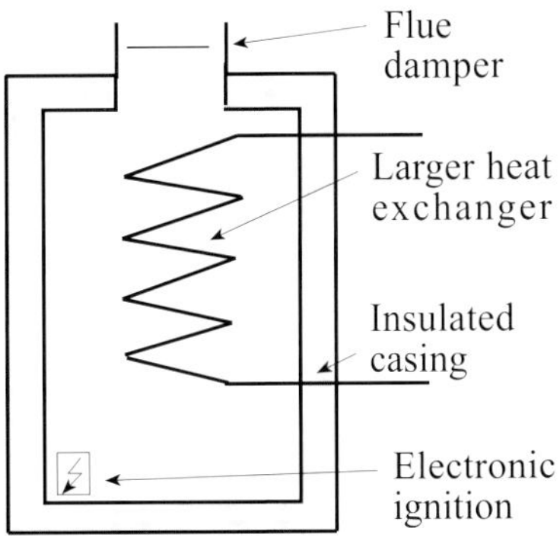

Figure 1.7 Features of a high efficiency boiler

To ensure that the boiler condenses the return water temperature must be below 53°C. Seasonal efficiencies are as high as 92%. This mode of operation does however, present design challenges. Firstly the cooled flue gases lose their buoyancy and are generally cleared by a fan (figure 1.8). Secondly the flue gas condensate is slightly acidic and so the heat exchanger must be made of none corrosive materials such as stainless steel. The condensate itself must be collected and drained away. All of these features add about 50% to the cost of a condensing boiler in comparison to a standard boiler for the same rating. However their high efficiency makes them economical with the extra capital costs typically being recovered in the value of energy savings within three years.

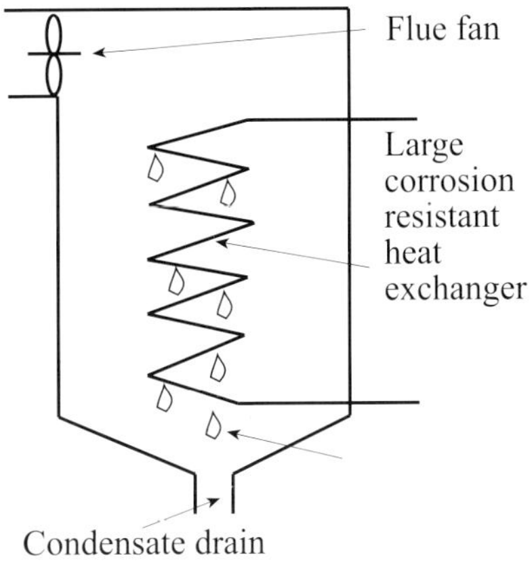

Figure 1.8 Features of a condensing boiler

HEATING
TOTAL
Wall Hung Boilers
Condensing Boilers
Atmospheric Boilers
High Efficiency Boilers
Pressure Jet Boilers
Direct Fired Water Heaters
Instantaneuos Water Heater
Calorifiers
Sealed System Pressurisation Sets
Boilers suitable for natural gas, LPG, oil or dual fuel
Flue Components
Packaged Fan Dilution Units
Flue System Design & Installation
the solution is HAMWORTHY
The specialists in design and manufacture of innovative commercial heating equipment
There's never been a wider choice from Hamworthy
Hamworthy
Hamworthy Heating Limited
Fleets Corner Poole BH17 0HH England
Tel: 01202 662552 Fax: 01202 665111
Email: sales@hamworthy-heating.com
www.hamworthy-heating.com

BOILER LOAD AND EFFICIENCY

The efficiency of a boiler varies with the load upon it. High load is when the boiler is being asked to do a great deal of heating. For example, first thing in the morning when the building and domestic hot water are both cold. In this situation the boiler will fire continuously and the flue and casing losses will be small when compared to the heat being input to the rooms. An example of a low load situation is at the end of the day when the building has warmed through and the tanks are filled with hot water. The boiler will be cycling, that is firing for short periods then stopping just to keep heat levels topped up. Almost as much heat will be lost by convection up the flue as is given to the heating system. Hence efficiency will be low. Figure 1.9 shows a graph of efficiency against proportion of full load for the three types of boiler discussed previously.

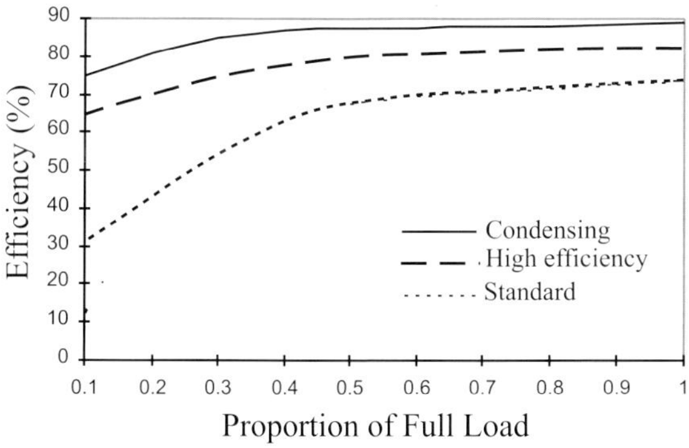

Figure 1.9 Graph of efficiency vs load

From the graph it can be seen that whilst standard boilers are effective when operated at high loads their efficiency falls off when the load on them decreases. The efficiency can fall as low as 35%. In comparison, the high efficiency boiler has a higher efficiency overall and has improved low load efficiencies. The condensing boiler has high efficiencies at all loads. The efficiency at low loads remains high at 75%. Higher efficiencies mean lower fuel costs and less pollution.

MULTIPLE BOILERS

In non-domestic buildings one way of ensuring that boilers fire near their high load rating is to operate them as part of a multiple system of boilers. This is recognised by the building regulations (L4) which require specific controls for heating systems over 100 kW rating. As an example, a multiple system of boilers used to satisfy a 100 kW load is shown in figure 1.10. It can be seen that the 100 kW load is provided by four 25 kW boilers feeding heated water into a common flow pipe and supplied by a common return. The first benefit of this arrangement of boilers is that there is back up if one of the boilers should fail. It can be isolated and heating can still be provided, albeit at a reduced capacity, by the other boilers. The second benefit is that the boilers are fired in a progressive manner to satisfy the load. So for example in the morning when there is a high load situation all of the boilers will fire. Later in the day when the building has started to warm through. Boilers 1 and 2 will fire continuously with boilers 3 and 4 shut down. At the end of the day when top up heating only is required only boiler 1 will be firing. The progressive mode of operation means that each boiler will only be firing near its full output rating. The system as a whole will therefore maintain a high efficiency even though the load is decreasing.

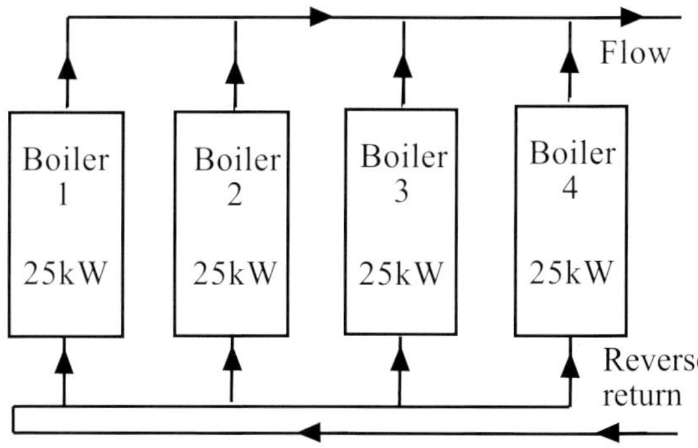

Figure 1.10 Multiple boilers for a 100 kW load

Progressive operation of the boilers requires a control process known as boiler step control. It is based on boiler flow temperature. If this falls it is an indication of increased demand for heating. As a result more boilers will be made to fire. Boiler 1 will be required to fire for more hours than any other boiler since it will operate during both high and low load situations. The boiler which is the first to fire up and last to switch off in any heating period is known as the lead boiler. To avoid unbalanced wear on the boilers the lead boiler will be cycled each week. So in week one boiler 1 will lead, in week two boiler 2 will lead and so on until after four weeks boiler one will once again be the lead boiler.

3

Keeping tabs on Energy Efficiency

Combined heat and power

Combined heat and power systems have the potential to greatly reduce primary energy consumption and carbon emissions by simultaneously providing heat and electricity at the point of use. Traditional electricity generation produces heat which is mostly wasted.

However, to be financially viable, there needs to be simultaneous demand for these services for a substantial part of the year. Leisure complexes with swimming pools and hospitals are typical applications. Where individual buildings do not provide a suitable load pattern, linking several buildings together may do – especially if they have different usage: offices and homes, for example. Community heating (district heating) is one example of this.

Usually a CHP unit is sized to provide the 'base load' of heat demand, with a conventional heating system providing additional heat at times of peak demand.

For a variety of technical, operational and financial reasons, the optimal design of CHP systems is rather more complex than for most other energy efficiency measures, and specialist advice will usually be needed.

Guidance on the implementation of CHP and related issues including site-specific advice is available from Action Energy.

Guidance documents cover such issues as:

- Good Practice Guides on implementation, operation and maintenance issues.
- Case Studies and Good Practice Guides for a range of building types.
- the use of CHP with community heating.

Further information

For further information on building-related energy efficiency, all it takes is one free phone call. Take action.

Call Action Energy today.
0800 58 57 94
www.actionenergy.org.uk

1.2.2 Combined Heat and Power

Combined heat and power (CHP) units are an additional source of heat for some buildings. These units are based on internal combustion engines similar to car or tractor engines (figure 1.11). They have spark plugs an engine block and cylinders. The first difference to a vehicle engine is that instead of running on petrol or diesel fuel they run on gas (natural, biogas or bottled). Secondly rather than drive a set of wheels the motive force generated by the engine is used to drive an electricity generator. This is the "power" part of the output. Instead of a radiator to exhaust the waste heat from the engine to atmosphere the CHP unit has a heat exchanger which transfers this heat into the heating system circulation. CHP units are also based on gas turbine technology giving increased heat and power outputs.

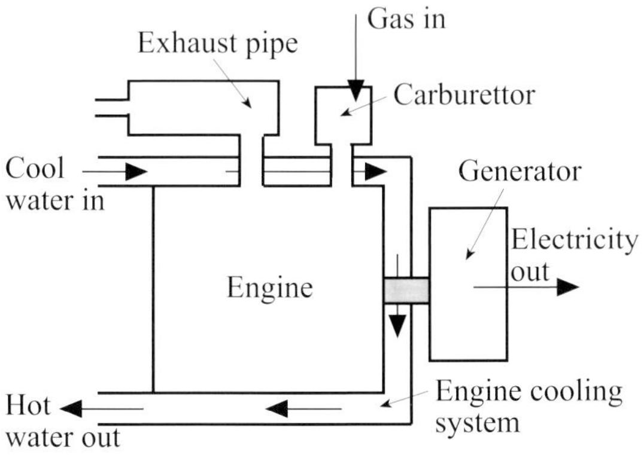

Figure 1.11 Combined heat and power unit

The efficiency of the CHP unit at producing heat is lower than that of a gas boiler at approximately 60%. This can be increased by 5% if an extra heat exchanger is used to recover additional heat from the exhaust pipe and oil cooler. This gives a heating efficiency which is still lower than a gas boiler. However, when the heat output is combined with the energy value of the electricity output the efficiency is greatly increased, to approximately 85%. This is illustrated in the energy flow diagram for the CHP unit shown in figure 1.12.

There are environmental benefits to be obtained from the use of CHP units. These are derived from replacing power station generated electricity by CHP generated electricity. For example, the efficiency of a coal fired power station at producing electricity is approximately 35%. 65% of the energy value of the coal is lost as waste heat in the flue and cooling towers of the power station and in transmission losses in the grid cables. A CHP unit has comparable electricity production efficiency but the waste heat is used in the buildings heating system.

The economics of CHP units is complex and involves a balance between savings in energy bills against capital and running costs. CHP units produce electricity at a unit cost which is much cheaper than can be purchased from the grid. But for the savings from this to pay back the capital cost of the installation the CHP unit must run for the maximum number of hours possible. Balanced against this is the fact that the CHP unit, like any engine, requires periodic routine maintenance. This involves changing oil, filters and spark plugs. CHP maintenance costs are high.

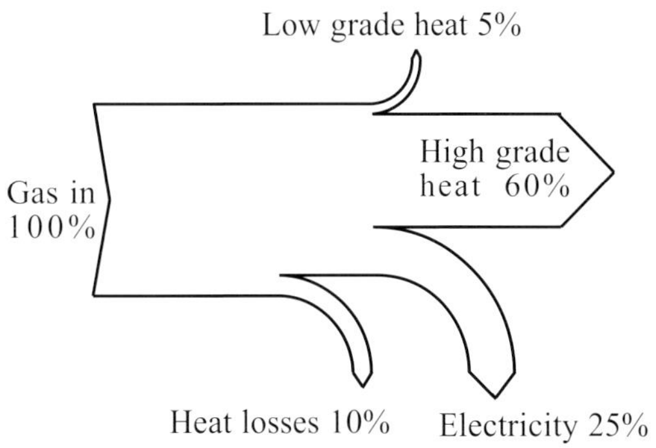

Figure 1.12 Energy flow diagram for a CHP unit

To maximise the running hours which is necessary to pay back the capital and maintenance costs the following three stranded strategy must be followed.

Firstly the CHP energy output must be matched to the building in which it is installed. A unit must be selected whose output satisfies the base heating and electricity demand which occurs all year round. If the CHP gives out more heat than is required the system controls will

Holistic low-energy design

For a building to consume the minimum amount of energy while maintaining acceptable levels of thermal comfort it must be designed holistically. All of the elements that have an impact on a building's energy consumption must be considered.

When considered holistically, passive solar strategies can provide approximately 10% of the space heating energy use of a typical dwelling. In commercial properties, such as offices, where heat gains occur from lighting, occupants and office equipment, care needs to be taken to avoid overheating especially during summer months.

Effective use of daylighting can greatly reduce electric lighting bills.

Element	Design issue
Siting	Exposed sites, solar/daylight access
Form	Surface area, volume, internal layout, orientation
Fabric	Insulation levels, sealing, workmanship
Ventilation	Uncontrolled infiltration, mechanical ventilation, heat recovery
Daylight	Window design, distribution, orientation, internal surfaces
Artificial light	Lighting design, efficiency of lamps/luminaires, control
Passive solar heat	Collection systems, thermal mass, distribution of heat
Mechanical heating	System selection, efficiency of components, control
Cooling	Passive cooling, system selection, efficiency of components, control
Services	Fans, pumps, lifts, renewables, etc
Post-occupancy	Monitoring and targeting, energy management

Factors in low-energy design

Further information

For further information on building-related energy efficiency, all it takes is one free phone call.
Take action.
Call Action Energy today.
0800 58 57 94
www.actionenergy.org.uk

switch it off to avoid over heating, cutting down the running hours. If the CHP gives out more electricity than is required by the building it will have to be used by other buildings on the site or exported to the grid. The export of electricity requires the installation of extra meters and unfortunately the price paid by the electricity companies for electricity deposited into the grid is low. It can be seen therefore that a high and consistent base demand is required for economic operation of the CHP unit. This tends to make them more suitable to buildings such as leisure centres and hotels with swimming pools.

Secondly, the CHP unit will be part of a heating system incorporating gas boilers to provide the above base heat demand. To make sure the CHP has maximum chance to run it must be the first heating device the system return water encounters on its way back to the plant room. In other words the CHP must be in series with the boiler heating system.

Thirdly, the CHP must undergo routine maintained at the specified intervals. In addition, many units are fitted with sensors, control devices and modems that allow them to auto dial a maintenance company if the CHP should stop running due to the occurrence of a fault. This will allow rapid attendance by a service engineer to rectify the fault.

1.3 Pumps

It is the job of the pumps to make the water circulate between the boiler and heat emitters within the heating system pipe work.

The three main components of a pump (figure 1.13) are an electric motor, an impeller and the casing. The electric motor is directly coupled to the drive shaft of the impeller. Water on the inlet side enters the pump in the centre of the impeller. The impeller rotates driving the water out towards the casing by centrifugal force. The water outlet is situated off the centre axis of the pump. As a result the pump casing must be cast to arrange the inlet and outlet flows to be along the same centre line. The pump is then known as an "in-line pump".

In domestic heating systems a single pump will suffice. However commercial heating systems contain a large volume of water which may have to be pumped great distances. In this situation high capacity twin head pumps are required. Twin pumps are required to give stand-by capacity if one of the pumps should fail. This is because a loss of pump power in a commercial building would result in an unacceptable loss of heating.

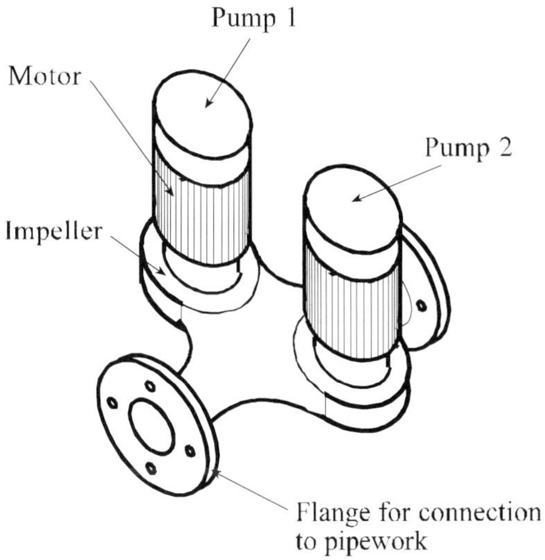

Figure 1.13 Twin head pump

Only one pump runs at a time, this is called the duty pump, the other acts as a stand-by. It is usual to run each pump for 1500 hours then change over to the other to even out the wear on them. This process can be carried out manually via the pump control panel. However in modern systems this is carried out automatically using a building energy management system (BEMS) (section 1.6). As well as routine cycling of pumps a BEMS can detect pumps failing if they are fitted with a suitable sensor and automatically isolate it and start up the stand by pump.

Variable speed pumps. An energy saving development in pump technology is the variable speed drive pump. This system does not run at a fixed speed but varies its speed and hence pumping power depending on the work it is required to do. For example, if a heating zone is warm enough valves will close isolating its heat emitters from the heating flow. As

IP3 - MOTORS.AND.DRIVES

Electric motors are everywhere in building services. They drive fans, pumps, lifts escalators and process machinery. In a typical prestige air conditioned office, fans and pumps account for 20% of the total electrical consumption (1). For comparison refrigeration only consumes 11% of the total. Motors are, therefore, key services components and major consumers of energy.

ELECTRIC MOTORS

Most electric motors used in building services are of the AC induction type. Single or three phase alternating current is fed through copper coils in the stator creating a magnetic field. This magnetic field induces another in the rotor. This causes the rotor to spin in the same way that like poles on bar magnets push apart. This spin can be used to drive the impellers of pumps (section 1.3) and fans (section 4.3).

Electric motors are everywhere in buildings. They are key components and major consumers of energy.

HIGH EFFICIENCY MOTORS

Motors are machines that convert electricity into movement. Like most machines their efficiency is less than 100%. The wasted proportion is seen as heat, arising from overcoming friction and created as a result of resistance in the windings of the motor. The efficiency of a typical 3kW motor is approximately 81%.

It is possible to increase the efficiency of motors by using low loss electrical steels and by increasing the thickness of wires used in the motor construction. This reduces resistive and inductive heating in the windings. As a result the motor cooling fan can be made smaller which adds to increased efficiency. These modifications increase the 81% motor efficiency by 4% i.e. up to 85%. This does not seem a large improvement but when you consider motors run for up to 24 hours a day the cumulative savings are very large. The additional cost of a high efficiency motor (about 25% more than a conventional motor) will typically be paid back within the first year of operation. Some manufacturers now offer high efficiency motors as standard

DRIVE SYSTEMS

The majority of motors run at a fixed speed. Variations in demand are usually satisfied using flow control devices. For example in a warm air heating system as the demand for heating falls the supply of warm air to the space will be reduced by closing a damper. The fan motor continues to operate at fixed speed.

There is a rule affecting motors, known as the cube law, which states that electricity savings are proportional to the cube of reductions in speed. This means that cutting the motor speed by 20% will give a 50% saving in electricity consumption. From this it can be seen that even modest reductions in motor speed will result in considerable energy savings. There are three types of variable speed drive (VSD). These are(2);

A two stage motor i.e. fast/slow/off operation. This is cheap and gives reasonable savings.
Electromechanical systems. Using gears, drive belts and slip disks to vary drive speed. These are robust but do not give maximum savings.
An inverter. This converts 50Hz mains electricity to DC. It then re converts it to AC at a frequency dependent on load. Increasing the frequency in response to increasing load increases the speed of the motor and vice versa. This system gives maximum flexibility and so maximises savings.

Additional benefits from VSDs are reduced maintenance costs and reductions in electrical standing charges.

Further information

1. Energy Efficiency Office. Energy Consumption Guide 19: Energy Efficiency in Offices. HMSO 1992.

2. Energy Efficiency Office. Good Practice Guide 2:Guidance notes for Reducing Energy Consumption Costs of Motor and Drive Systems. HMSO 1993.

a result less water will need to be pumped around the heating circuit. A variable speed pump will sense this and slow down. This is illustrated in Figure 1.14 which shows that the energy consumption of a fixed speed pump remains constant as the demand for water flow falls. The variable speed pump slows down to match demand resulting in a fourfold reduction in electricity consumption for each halving of pump speed.

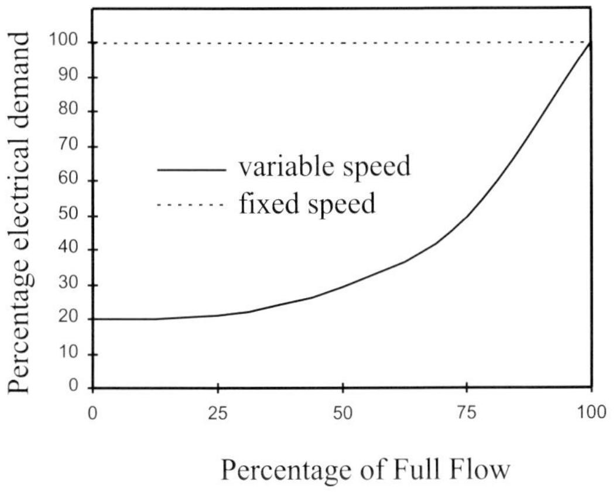

Figure 1.14 Graph of pump electrical use against percentage of flow

By exercising variable speed control of the pumps considerable amounts of energy and hence money can be saved. Using less energy also gives a reduction in the output of pollutant gasses from power stations (see IP3).

1.4 Heat Emitters

Heat emitters transfer the heat from the heating system to the rooms requiring warming. This is usually carried out by convection and radiation from a surface heated by the hot water flowing through the heating circuit. To avoid overheating the room some method of control is required. This is usually achieved by restricting the flow of heated water into the heat emitter using a valve.

In most domestic buildings heat is emitted to the rooms using radiators. Water heated to 80°C by the boiler flows into the radiator, raising its temperature. The radiator warms the room by losing heat to it. The radiator gives out heat partly by radiation but mainly by convection. Convection occurs when the radiator heats up the room air in contact with it. The air becomes less dense and so rises to the ceiling where it mixes with the rest of the room air. Cooler air from beneath the radiator is drawn up to repeat the process. Because it has lost heat to the room, water leaving the radiator and returning to the boiler is typically 10°C cooler than the flow temperature. In order to achieve this temperature drop the flow of water through all the radiators in the system must be regulated during commissioning. This is carried out by opening or closing lock shield valves fitted on the radiator outlets to increase or decrease the flow rate respectively.

For a given boiler flow temperature the heat output of a radiator is determined by the size of its surface area for convection. The single panel radiator (figure 1.15) is the simplest pattern. It is two pressed steel panels sealed by welding on all sides. Its shape gives it an internal void, which fills with hot water, and surface convolutions which increase its surface area. The single convector radiator has this same basic panel but has an additional corrugated plate spot welded to its back surface. This plate increases the effective surface area over which convective heat loss can take place. Two other patterns are shown which also increase heat output by increasing surface area further. The heat outputs of each of these radiators is given for a 600 mm high by 1000 mm long radiator. It can be seen that in comparison to the single panel radiator the heat outputs of the single convector, single panel / single convector radiators and double convector radiators are 43%, 112% and 170% bigger respectively.

The benefit of increased heat output is that the physical size of the radiator can be reduced for a given heat output. A feature which is useful in confined spaces such as where sill heights are low. However it must be remembered that the cost and depth of the radiators also increase with increasing heat output

Radiators are usually positioned beneath windows. This is a useful location as it is unlikely that furniture will be positioned here and also the heat output of the radiator will counteract the cold down draughts from the glazing. Radiators are rated in terms of their heat output which should be matched to the peak heat loss

I P 4 - SIZING.BOILERS.AND.HEAT.EMITTERS

Manufacturers produce a range of boiler and heat emitter sizes to satisfy the needs of various buildings. Before you can buy a boiler and connect it up to the heat emitters you need to know how much heat is required by each room which in turn informs you of the size of the boiler. Over sized boilers should be avoided since they will rarely operate at peak load and so will have low efficiencies. Under sized boilers will not give the required output and so room temperatures cannot be maintained against low outside temperatures.

Undersizing of boilers means temperatures cannot be maintained, oversizing results in low efficiencies.

HEAT EMITTER SIZING

Heat Emitters must be sized to supply the peak heating demand of a particular room. This is determined by calculating the peak fabric and ventilation heat loss rates of the room. Examples of how to do this are given in building science text books. The ambient conditions assumed for the calculation use -1°C for the outside air temperature along with the design indoor temperature. For a room occupied by people engaged in a low level of physical activity, such as an office or living room, the indoor temperature is assumed to be 21°C. The air change rate and fabric thermal properties are also required

The outcome of the heat loss rate calculation described above for a living room might give a heat loss rate of 2000Watts. This means that when it is 21°C inside and -1°C outside the room will be losing heat at a rate of 2000W. To maintain the internal temperature heat must be supplied to the room at the same rate. This is anologous to water pouring out of a hole in a bucket. To maintain the required water level (21°C) water must be poured into the bucket at the same speed at which it is leaving through the hole (the heat loss rate). For this reason the heat emitter for our example room should be sized at 2000W. Trade literature for heat emitters gives a range of useful information such as dimensions and mounting details. It also gives information on heat outputs. A suitable heat emitter can be chosen from these tables.

One difficulty of heating large rooms is to get adequate heat distribution throughout the room. Unless some kind of forced convection system is used, heat tends to be concentrated near the heat emitter. One way of achieving better distribution is to divide the heat input into the space using two or more heat emitters distributed evenly through the room.

BOILER SIZING

The process used to determine the fabric and ventilation heat loss rate for the individual room must be repeated for all rooms. If an indirect heating system is being used. The source of heat, usually a boiler, must be able to supply the total heating requirement of the heat emitters in all the rooms. For the four roomed house shown in figure IP4 it can be seen that the boiler power needs to be 3.3kW. If hot water is to be derived from this boiler an allowance (typically 3kW) must also be added for this purpose.

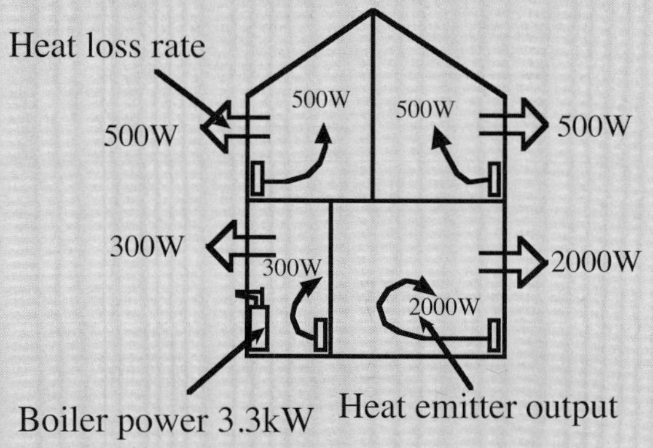

Figure IP4. Heat losses from room counteracted by appropriately sized heat emitters

In large buildings where the heat output of the boilers is measured in hundreds of kilowatts a multiple system of boilers must be used (section 1.2.1) to maintain high operating efficiencies.

rate of the room in which they are situated (see IP4). In large rooms the radiator output should be split and more than one radiator used. This will distribute the heat more evenly throughout the room.

Single panel - heat output 900W

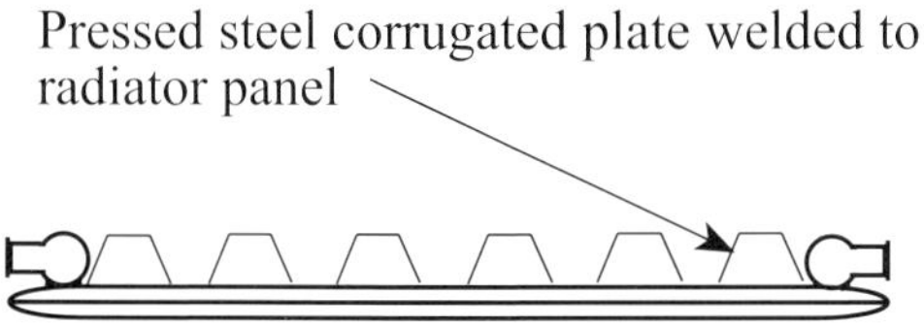

Single convector - heat output 1300W

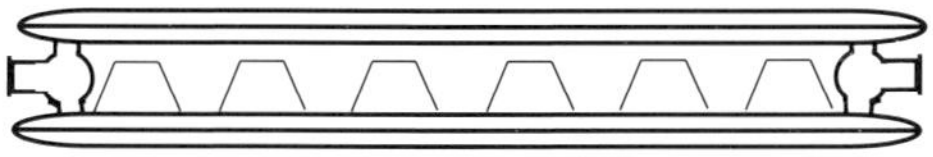

Double panel with single convector - heat output 1900W

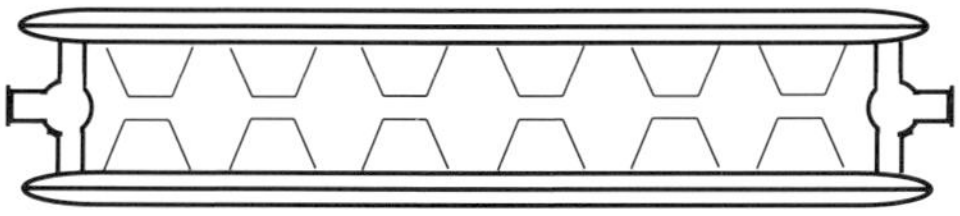

Double convector - heat output 2400W

Figure 1.15 Radiator patterns - plan views

COMMERCIAL HEAT EMITTERS.

As with domestic buildings, radiators are used in commercial buildings but in addition, a wide range of other heat emitters are encountered.

Low Surface Temperature (LST) Radiator. This is a radiator which is encased to prevent touching of the hot surfaces (figure 1.16). A top grille allows heat to leave the unit. LST radiators are suitable where high surface temperatures could cause burning. Examples are aged persons homes or nursery schools.

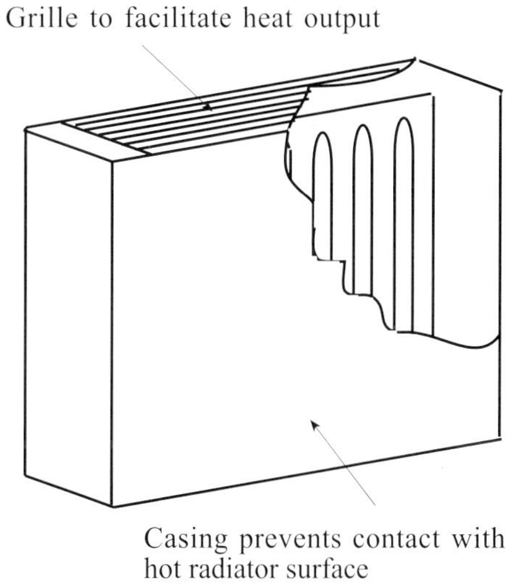

Figure 1.16 Low surface temperature radiator

Perimeter Radiator. This radiator is constructed from a tube which has had fins added to increase its surface area for heat output (figure 1.17).

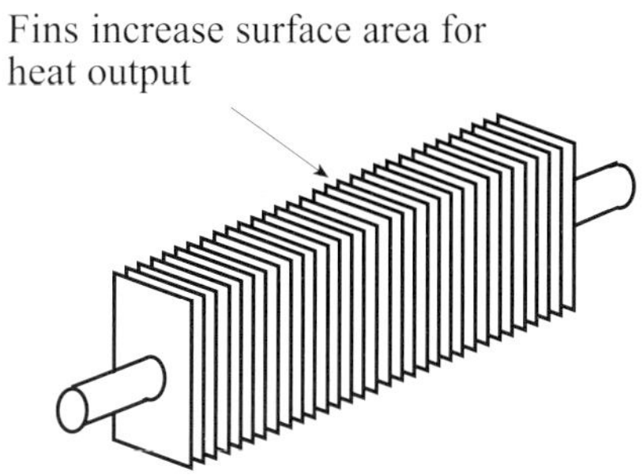

Figure 1.17 Perimeter radiator

NOTES

The unit may only be 150mm high but it is long in length. Perimeter radiators are typically used along the entire outer edge of highly glazed spaces. Here their heat output counteracts cold down draughts from the glazing. Its low height makes it unobtrusive. A modification of perimeter heating is to recess the radiator into the floor depth and cover it with a grille to form a perimeter convector heater.

Convector Heater. Convector heaters are constructed from a cabinet in which there is a finned coil heated by water flowing through it from the heating system (figure 1.18). Air inside the casing is heated by contact with the heating coil causing it to rise up through the convector and out of the upper grille to heat the room. The convection current carries on this cycle by drawing cool room air into the cabinet via the lower grille. A filter behind the inlet grille removes dust from the airstream.

Figure 1.18 Convector heater

The heat output of the unit can be increased and the time taken to heat the room reduced by fitting a fan into the casing to drive the circulation of air through the heater. The heater is then known as a fan convector. Fan noise can be a problem in some quiet locations but heat output can be regulated more effectively by switching the fan on and off as required.

Radiant Panel. These heat emitters are composed of copper tubes welded onto metal plates (figure 1.19). Flexible connectors are then used to connect a series of these plates together. The panels, which in offices are perforated and painted, are hung to form part of the suspended ceiling. Water from the heating system is passed through the tubing causing the temperature of the panels to increase. The space below is then heated by convection and radiation.

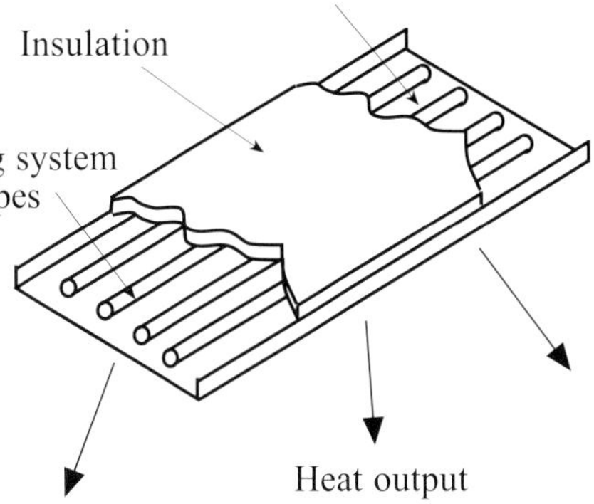

Figure 1.19 Radiant panel

The advantages of radiant panels is that they do not take up any wall space and their heat output will not be affected by furniture such as filing cabinets or desks pushed up against walls.

Underfloor Heating. This method utilises the entire floor, and sometimes the walls, of a room as a heat emitter (figure 1.20). Water from the heating system is passed through polymer pipes embedded in the floor screed. The flow temperature, at approximately 24°C, is much lower than for other types of heat emitter. This is possible because of the large surface area created by utilising the floor for heat output. This is in contrast to the smaller but hotter surface area of a radiator.

There are many benefits from using an underfloor heating system. These are;

- Wall space is not taken up by heat emitters.

- large spaces which are difficult to heat evenly from perimeter heat emitters can be uniformly heated.

I P 5 - H U M A N . T H E R M A L . C O M F O R T

Human thermal comfort is determined by the way individuals perceive the temperature of their environment i.e is it too hot or too cold. This perception depends on personal preferences. As a result, within a group of people in the same room, some will feel comfortable, some too hot and some too cold. Building professionals must use their knowledge of the building fabric, heating services and human physiology to ensure that the majority of people in a space are satisfied with the temperature. There are some serious health concerns in buildings (see IP10). But lack of thermal comfort is a chronic problem which affects many people in badly designed buildings.

Lack of thermal comfort is a chronic problem which affects many people in badly designed or serviced buildings

THERMAL COMFORT

To be comfortable a person requires a stable core body temperature of 37.5°C. To achieve stability any heat inputs to the body must be balanced by a heat output. Extra heat input or reduced heat losses will cause the subject to feel warmer. Extra heat loss or reduced heat gains causes the subject to feel colder. Heat gains to and losses from the body are illustrated in figure IP5. **Convective heat** gains and losses are created when warm air moves into or out of contact with the body respectively. Convective heat transfers are strongly dependent on air movement around the body. **Conductive heat** gains and losses occur due to body contact with hot or cold surfaces respectively. Since normal contact with room surfaces is restricted to the soles of the feet this does not constitute a major component. **Radiative heat** gains and losses occur when a person is positioned next to a warm or cold surface respectively. The human body is very sensitive to radiant energy and so this component has a strong affect on comfort. In addition to the three basic forms of heat transfer there are two others related to the human body. The first is heat gain by the body due to metabolism. The body burns food to grow, repair itself and cause movement. A by-product is heat. The amount of heat gained by the body is substantial and increases with the level of activity. This is illustrated in the following table.

Activity	Metabolic Heat Output (W)
Sedentary	100
Active (light work)	150
Very Active	250

Secondly, additional heat loss arises due to evaporation of moisture from the lungs and skin. Latent heat is absorbed (see IP6) which cools the body. This cooling effect is increased in dry (low RH) environments. In high relative humidity environments evaporation is suppressed. The space is then commonly referred to as being hot and "humid".

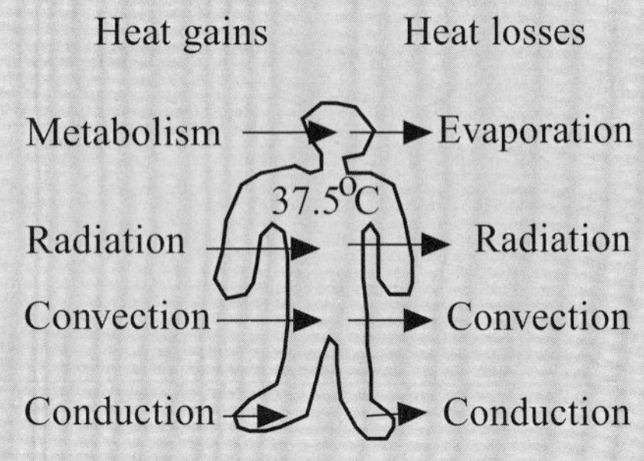

Figure IP5 Balance of body heat gains and losses

FACTORS AFFECTING THERMAL COMFORT

Anything which changes the balance of heat inputs and outputs will affect thermal comfort. For example, if air temperatures rise heat gains will increase. Turning on a fan in response causes air movement increasing heat losses. This returns the body to thermal balance and comfort. The body itself is very effective at thermoregulation e.g raising hairs for insulation, shivering for metabolic heating, variable skin blood flow to regulate body heat loss and sweating to cool evaporatively. Other variables are;

Amount of clothing	(insulation)
Temperature gradients	(differential losses)
Average surface temperature	(radiant transfers)
Relative humidity	(evaporation)

• Thermal gradients decrease from foot to head improving thermal comfort and reducing the risk of stratification.

• The low flow temperatures utilised in underfloor heating makes them ideal for use with condensing boilers (section 1.2.1). The low return temperatures will increase the tendency of the boiler to operate in condensing mode.

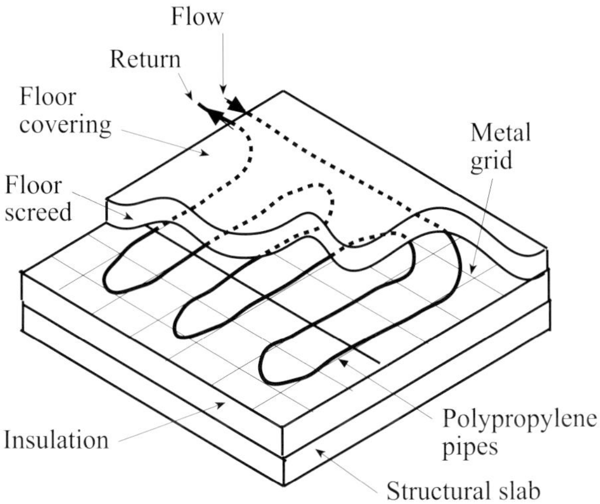

Figure 1.20 Underfloor heating system

1.5 Domestic Hot Water

Hot water is obtained from domestic central heating systems in two ways. Firstly, by the use of indirect cylinders which draw some of the heat away from the heating circuit to provide hot water. The second is to use combination ("combi") boilers which generate hot water instantaneously as it is required.

Indirect Cylinder. Some of the heated water flowing through the heating circuit is diverted, using a three way valve, through a calorifier within the hot water cylinder (figure 1.21). The calorifier is essentially a coil of copper tube through which the heating system water flows. Heat transfer from the calorifier warms up the cold water held in the cylinder. The heated water rises to the top of the tank where it is drawn off to the taps. The cold feed, which enters at the base of the cylinder, is from a mains fed tank which is often built in to the top of the cylinder to form an integral unit. Since the feed water tank is open to the air the system is referred to as a vented or non sealed system. The benefits of indirect water heating are that the central heating boiler performs two functions (space and water heating) and that there is a stored volume of hot water ready to meet peak demands.

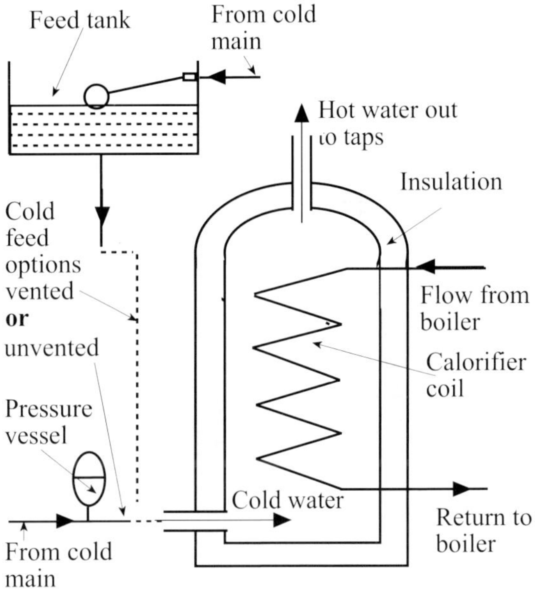

Figure 1.21 Indirect dhw cylinder

Unvented indirect cylinders are similar to the system described above in that a calorifier fed from the wet heating system is used to heat a stored volume of water. The difference is that the cold feed is from a direct connection to the cold main instead of from a tank. As a result hot water from the cylinder is fed to the taps at mains pressure. This gives a greater flow rate than a tank fed system. Since there is no opening in the system to the air it is known as a sealed or unvented system. Two of the benefits of this systems are that showers can be successfully fed from them and that the plumbing necessary for a feed tank is not required.

When water is heated it expands. The increase in pressure caused by this would damage unvented system since the pressure cannot be released. Because of this unvented hot water storage systems are fitted with a small expansion vessel to take up the

IP6-THERMAL.CAPACITY,SENSIBLE.AND.LATENT.HEAT

THERMAL CAPACITY

Thermal capacity is a measure of the ability of a material to absorb heat. It is usually specified in terms of the specific heat capacity of the material. This is the amount of heat, measured in Joules (see IP2), that one kilogram of the material must absorb to raise its temperature by 1°C. The units of specific heat capacity are J/kg/°C.

For example the specific heat capacity of water is 4200 J/kg/°C, of air is 993J/kg/°C and of stone is 3300J/kg/°C. It can be seen that per kilogram stone has a much greater heat carrying capacity than air and that water has a higher heat carrying capacity than stone. This has consequences for the building services industry. Air cannot carry as much heat per unit volume as water. As a consequence heat distribution systems which use air must be much larger than hydronic distribution systems to carry the greater volumes required. The relatively high thermal capacity of dense materials such as stone is used for thermal storage. One example is the use of special blocks in electric storage heaters (section3.1).

SENSIBLE AND LATENT HEAT

Sensible heat and latent heat are both forms of thermal energy. The difference in name arises as a result of what happens to a material when the thermal energy is being absorbed.

The absorption of **Latent heat** causes a change of state. One example is the absorption of the latent heat of vaporisation by water to change it from a liquid to a gas (water vapour). It should be noted that a substance gives out latent heat when the phase change is reversed. For example the latent heat absorbed by a refrigerant in the evaporator coil of a vapour compression chiller is released once more when the refrigerant condenses in the condenser.

Absorption of **sensible heat** causes an increase in the temperature of the object. The amount by which the temperature rises depends on the amount of energy absorbed, the mass of the material and its specific heat capacity (see above). Sensible energy is released by the object as it cools.
Sensible and latent heat are best illustrated using water as an example.

Figure IP6 shows what happens if a 1kg block of ice at 0°C is placed in a beaker over a Bunsen burner. The ice absorbs heat from the flame but its temperature does not increase instead it changes state, it begins to melt. The heat absorbed is called the latent heat of fusion (units J/kg). When completely melted, further heat input causes the temperature of the water to rise. The thermal energy now being absorbed is called sensible heat. The temperature rise continues until 100°C is reached. At this point the temperature once again stabilises and a second change of state occurs. This time from liquid to vapour. The heat absorbed is called the latent heat of vaporisation. This continues until all the liquid is converted to vapour.

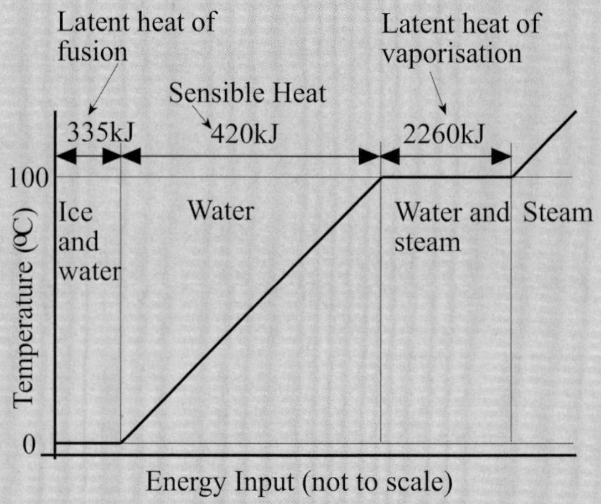

Figure IP6: Heating of ice

Consideration of figure IP6 shows some interesting features. The first is that substantially more energy is required for the phase change from water at 100°C to steam than for the heating from 0 to 100°C. This means that steam at 100°C contains far more energy than water at this temperature. This is why steam is a useful heat transfer medium. It contains a lot of energy so distribution pipes can be kept small whilst transferring large amounts of heat to the heat emitters. It is also dangerous. If steam escapes and condenses onto human skin all of the latent heat of vaporisation is re released which can cause severe burns.

extra volume of water created by heating.

Energy issues. In summer the low loads encountered by a central heating boiler required to generate hot water only leads to reduced boiler efficiency. It is therefore recommended that indirect cylinders are used in conjunction with a condensing boiler (section 1.2.1). The cylinder itself must be well insulated to reduce heat loss from the stored hot water. These heat losses are known as standing heat losses. The alternative approach to energy efficiency in commercial buildings is to use a stand alone direct water heater (section 3.3)

Combi Boiler. A combi boiler is a combustion device which has two heat exchangers, one for the space heating system and one for the domestic hot water (figure 1.22). Cold water is fed into the unit directly from the mains. Turning on the hot tap allows cold mains water to flow through the boiler. The pressure changes cause the burner to fire. Hot water is therefore generated as needed. There is little or no stored volume of water. The casing of the boiler also houses pumps, controls and a pressure vessel. A pressure vessel is required since the system is sealed and provision has to be made for the increase in pressure that occurs as water in the system is heated.

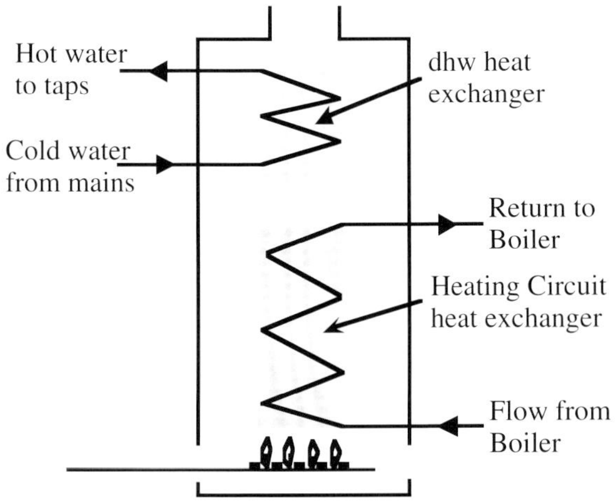

Figure 1.22 Combination boiler

One of the major benefits of this system is that it is simpler and less costly to install as a feed and expansion tank, indirect cylinder or cold water storage tank are required. This can be seen by comparing figure 1.23 which shows a combination boiler heating system with figure 1.1 which shows an open vented heating system with indirect cylinder.

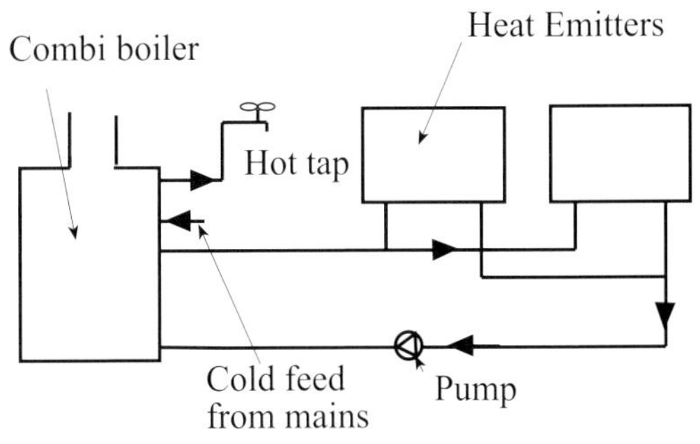

Figure 1.23 Combination boiler heating system

DHW FOR COMMERCIAL BUILDINGS.

Hot water in commercial buildings can be provided by indirect cylinders as discussed previously. The cylinders are however much bigger due to the increased demand for hot water experienced in larger buildings and are often referred to as calorifiers. Another method of dhw production is to replace the indirect cylinder with a water to water plate heat exchanger. The source of heat is still the indirect heating system.

Water to water plate heat exchangers are built of a sandwich of convoluted thin plates (figures 1.24 and 1.25). Alternate voids created by the plates carry heating circuit and dhw flows respectively. Heat is transferred from the heating circuit flow to the dhw flow by conduction across the thin metal separating the two flows. This process is fast enough to produce hot water instantaneously.

Plate heat exchangers have a number of advantages over storage systems;

- There are no standing heat losses since there is no stored volume of water to cool down overnight or at weekends.

- Legionnaires disease can arise where water is allowed to stand at the incubation temperature of the legionella bacteria. Since standing water is eliminated

IP 7/1 - PLANT.ROOM.POSITION.AND.SIZE

Building services can represent 50% of the cost of a highly serviced building and take up to 30% of its floor area. This brings the importance of building services in the construction process clearly into view. It also illustrates the need for early allocation of space for and planning of building services. The main elements involved are plantrooms and horizontal and vertical service runs. Space allocation must take into consideration the need for adequate space and access for servicing and, if it is felt necessary, provision for flexibility and future developments. Service runs permeate throughout the entire height, length and width of a building therefore, any building services designs must be made with due regard for the structure. Integration of services with the structure is an important element in the building design process.

Building Services can represent 50% of the cost of a highly serviced building and take up 30% of its floor area

This information panel does not have enough space to cover this subject in depth, but three main issues relating to the planning of services can be highlighted. These are space, location and distribution.

SPACE FOR SERVICES

There are many types of building and each one will have a different servicing requirement. Even within similar building types there are a range of solutions available. One example is office buildings that can be fully air conditioned, naturally ventilated or operate with both systems (mixed mode). The obvious rule is that the greater the need for building services the greater is the need for space to accommodate them. So for example a simple naturally ventilated heated office will devote 4-5% of its total floor space to plant whereas for a speculative air conditioned office this will rise to 6-9%. Highly serviced buildings such as sports centres with leisure pools may need to allocate 15-30% of the total floor area to services. For an individual building the final determinant of space requirement is the load on the heating or air conditioning system. Reducing the need for heating or air conditioning by using low energy design principles will cut down on the need for energy consuming plant and therefore on the space required to accommodate it.

At the early design stage rules of thumb will be sufficient to make an initial allocation of space. (see BSRIA Technical Note TN 17/95: Rules of Thumb, BSRIA 1995). This figure can be refined at a later stage when exact details are known.

Having arrived at a figure the space can be concentrated in one place which is usual for small to medium sized buildings. The possibility of dividing the space up and spreading it through the building depends on the layout of the building. If the building covers a large area then it may be economical to have smaller but more numerous plantrooms distributed throughout the site each satisfying individual zones or sections of the building. The advantage is that distribution runs are kept short and pipe and duct diameters can be smaller to reflect the reduced floor areas served.

LOCATION

Plantrooms can be located anywhere in the building but noise considerations, weight of equipment to be accommodated and ease of access for maintenance means that plantrooms containing heavy equipment such as boilers and chillers tend to be located on the ground floor or basement. However, modern low water content boilers (e.g. Hamworthy Wessex)are designed to be light for roof top installation. Air handling units are lightweight but bulky. This means they can be accommodated on rooftops where they are not taking up lettable space and structural requirements are not critical. The rooftop is a useful location for taking in air which is generally fresher than at ground level. The rooftop also gives the cooling system condenser access to the outside air for waste heat rejection.

Continued on page 32

with plate heat exchangers the possibility of infection is avoided.

- Plate heat exchangers are physically much smaller than indirect cylinders. This makes them useful where space is limited.

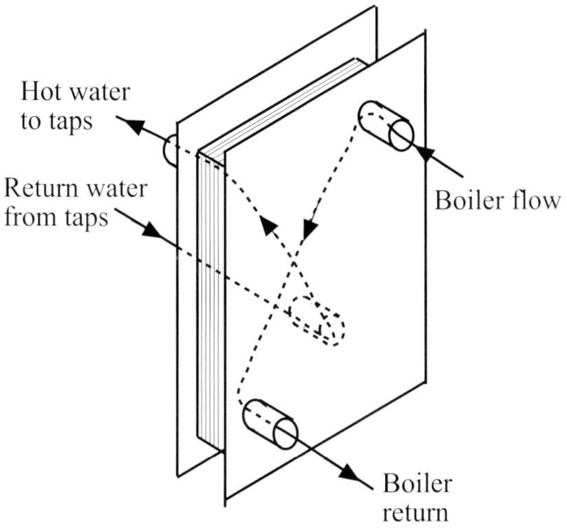

Figure 1.24 Water to water plate heat exchanger

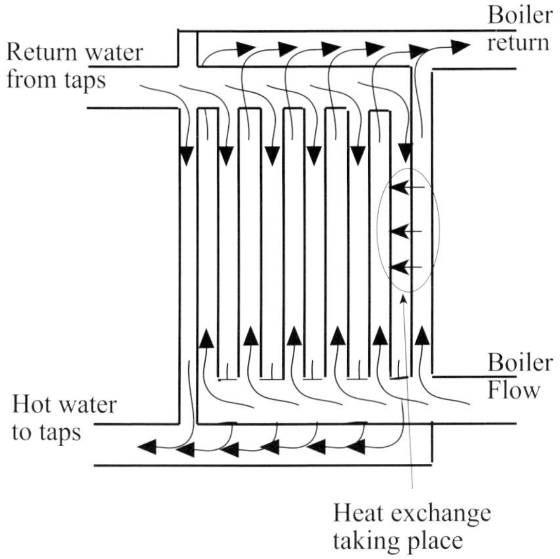

Figure 1.25 Cross sectional diagram of a plate heat exchanger

Since there is no stored hot water, which is used to satisfy demand at times of peak usage, the heat exchanger must be sized to satisfy the peak hot water demand of the building.

DHW DISTRIBUTION

A significant difference between domestic and commercial dhw systems is the way in which the hot water is distributed to the taps. In domestic properties a single pipe directs water from the cylinder to the tap. In combi systems the energy to do this comes from mains pressure. In indirect cylinder systems the water is moved by gravity. Both of these mechanisms are adequate because the pipe lengths are small. In large buildings, however, the pipe lengths become longer due to the large distances between hot water production and use. Standing heat losses from these long pipes would cool the water in them and result in tepid water being drawn off from remote taps. Running off this tepid water until hot water was obtained and the heat lost from the pipes themselves would result in wasted energy. To ensure that water is always available at the tap they are usually supplied from a secondary hot water loop (figure 1.26). A secondary dhw pump continually circulates hot water from the cylinder or heat exchanger around this circuit. As a result hot water is always available at the taps. Pipe insulation ensures heat lost from the pipes is minimised (see IP1).

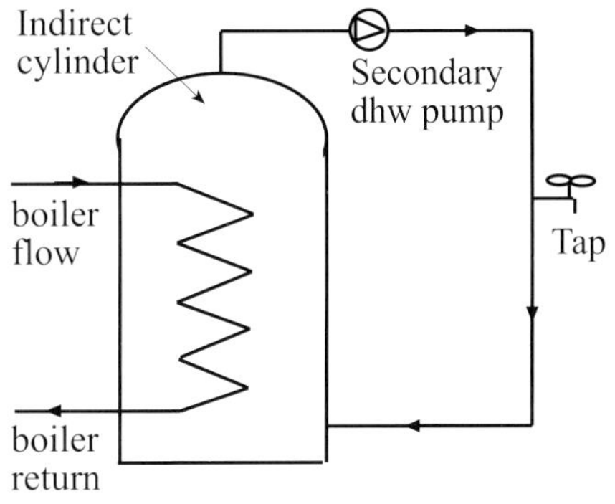

Figure 1.26 Secondary dhw circuit

IP7/2 - PLANT.ROOM.POSITION.AND.SIZE

From page 30

DISTRIBUTION

There are a number of services which may require distribution throughout a building. These include; hot and chilled water, potable water, electrical power and lighting, control cabling, conditioned air, communications cables and fire systems cabling. Taking heating as an example hot water is generated in the plantroom by the boilers. It must then be distributed to the heat emitters in each room. Finally, water which has had its heat removed must be returned to the plantroom for re heating. It can be seen that vertical runs of pipe are required to carry heated water to each floor. Horizontal runs of pipe are required to distribute the hot water to each heat emitter. In domestic buildings vertical pipes are surface mounted and boxed in for cover. Horizontal runs are made between joists and under the floorboards. In commercial buildings vertical service shafts are required. Horizontal distribution usually takes place under a raised floor system or above a suspended ceiling.

Ducting is the most difficult system to accommodate since it has a much larger cross sectional area than water pipes. This is especially so close to the air handling unit where the ducting must carry all the conditioned air for each space. The cross sectional area reduces the further away from the air handling unit you are as the conditioned air is progressively divided off into successive spaces.

Service runs should preferably be linear. This provides economy of installation and operation. A change of direction in ductiong or pipework requires additional components and fabrication. Bends and junctions offer greater resistence to fluid flow. As a result a larger and therefore greater energy consuming pump or fan would be required.

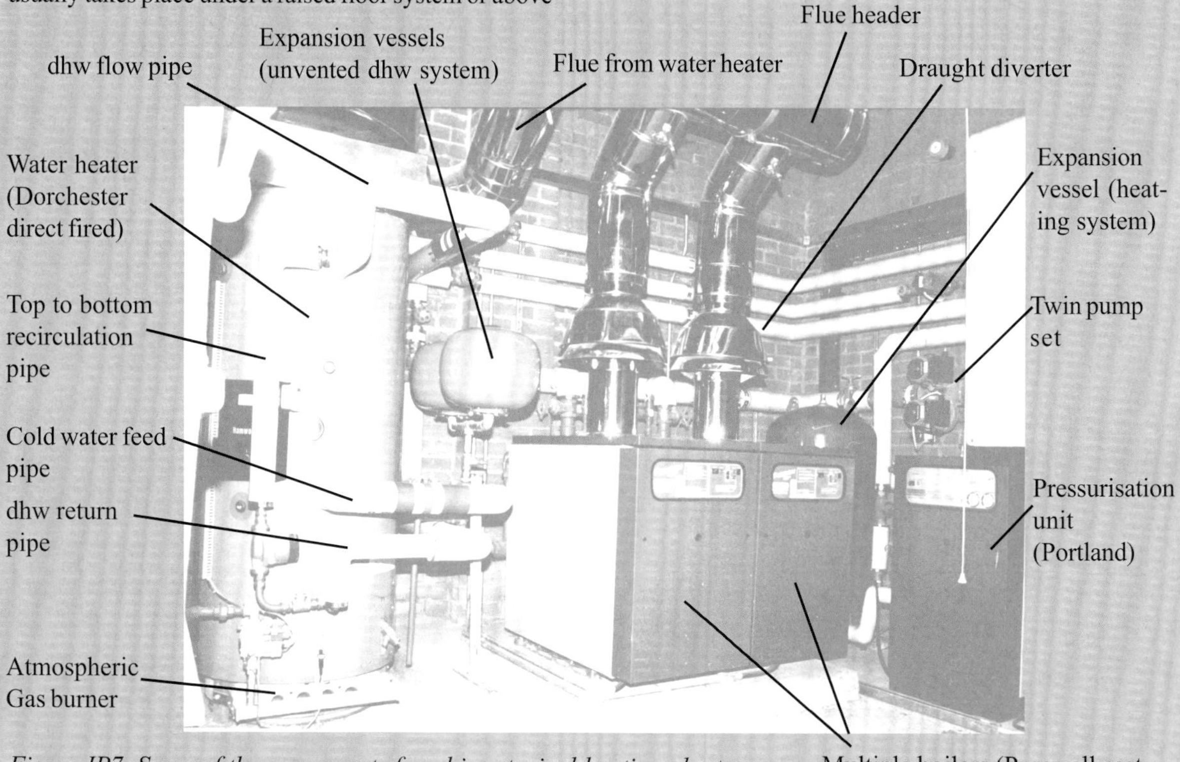

Figure IP7. Some of the components found in a typical heating plantroom photo: ***Hamworthy Heating Ltd.***

1.6 Controls

Controls are required to ensure that the heating system operates safely and efficiently and provides comfort for the building occupants. Figure 1.27 shows a typical arrangement of controls for a domestic central heating system. It is comprised of the following components.

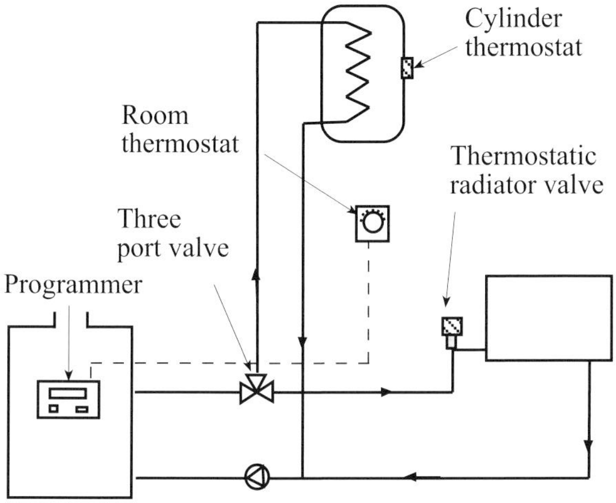

Figure 1.27 Domestic heating controls

Room Thermostat. Is a device which controls room temperatures. Control is made in relation to a preferred temperature setting made on the thermostat by the occupant. The thermostat is in fact a switch opened and closed as the room temperature rises above or falls below the temperature setting respectively.

The thermostat should be positioned in a representative room such as the living room at standing chest height away from sources of heat such as direct sunlight. This means it will accurately sense the air temperature experienced by an occupant in the room. When the thermostat switch is closed, current can flow through it. This is interpreted by the boiler as a call for heat. The boiler will fire, the pump will run and the three way valve will direct hot water to the radiators. When the room temperature rises above the preferred temperature setting, changes within the thermostat either electronically sensed or due to the differential expansion of metal in a bi metallic strip cause the thermostat switch to open. As a result the control current will stop and the boiler and the pump will switch off. It can be seen therefore that room temperatures are controlled by stopping and starting the flow of heat into the room as required.

Programmer. This is a time switch that determines the times within which the heating will respond to a call for heat from the room thermostat. The start and stop times between which the heating will be allowed to operate are entered into the programmer. For example, heating may be required from 07:00 to 08:30 in the morning then 17:00 to 23:30 in the evening. Modern microprocessor controlled programmers allow multiple daily heating periods and the ability to programme each day of the week with a different heating programme. An example is that the first "on" period at the weekend may start at 8:00 and end at 12:00 to reflect the fact that the occupant is not in work on that day.

Cylinder thermostat. This is a temperature controlled switch similar to the room thermostat. The difference is that it is clamped on to the indirect cylinder and therefore senses and controls the temperature of the dhw. When the switch is closed and therefore calling for heat, the three way valve will be instructed to divert boiler flow through the calorifier in the cylinder. This will cause the temperature of the water in the cylinder to rise. When the temperature reaches the setting on the thermostat the switch will open. The three way valve will then direct the flow away from the cylinder and back to the heating circuit.

Thermostatic Radiator Valve (TRV). This is a valve (section 1.7) fitted to the inlet of the radiator. Gas in the TRV head (figure 1.28) expands with temperature and pushes a gate downwards blocking the inlet flow. This will restrict or even stop the flow of heat into the radiator. The heat output will then be reduced

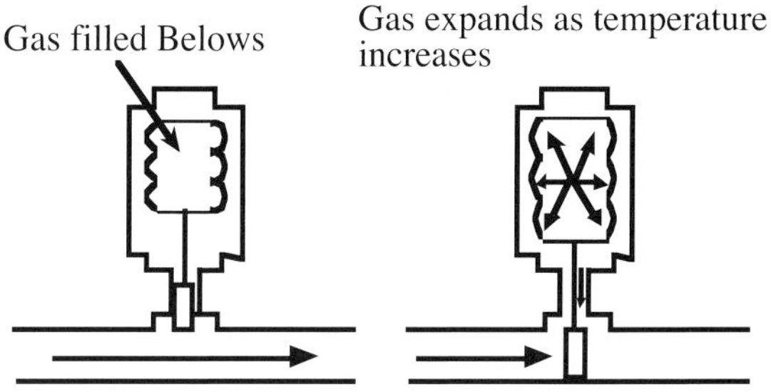

Figure 1.28 Thermostatic radiator valve

causing the room to begin to cool. As it does so the gas in the TRV head will contract and the valve will open up once more. TRV's allow room by room control of temperatures to be achieved. They are particularly useful in south facing rooms and rooms subject to casual heat gains, giving an extra layer of control beyond the single room thermostat.

CONTROLS FOR COMMERCIAL BUILDINGS

Large buildings cannot be controlled effectively using domestic control systems. The components must be scaled up and certain refinements made to provide adequate control. The difficulties encountered in heating control in large buildings arise due to their thermal sluggishness, different heat loads/gains and differences in hours of usage of spaces. Both time and temperature control require consideration.

Optimum start controller. Time control in domestic buildings is adequately carried out using the fixed on/off time controller described previously. This is because even on cold days the time taken for the majority of domestic buildings to warm up to comfortable temperatures is unlikely to be more than thirty minutes. This period when the building is warming up to the occupancy temperature setting is known as the pre heat period. In large buildings the pre heat period will be considerably longer due to the thermal inertia of the structure of the building and the heating system itself. The pre-heat period is also variable. It is longer in winter than in spring and autumn because the building cools more during the night. This means a fixed start and stop time would be wasteful. This is illustrated in figure 1.29. It can be seen that on a cold night the heating must come on at 1.0 a.m. to heat the building up to adequate levels by the start of occupancy at 8.00 am. If this fixed on time is retained on a mild night then the building is raised to the occupancy temperature at 5:30 am. This is two and a half hours prior to occupancy and so is wasteful.

To overcome this problem an optimum start controller is used. This is a device into which the operator inputs the times of the beginning and end of occupancy, say 8.00 am and 5.00 pm. The optimum start controller then monitors inside and outside temperatures, combines this with a knowledge of the thermal inertia of the building and as a result determines at which time to activate the heating in the morning to achieve the desired internal temperatures by the start of occupancy. In the above example the optimum start controller would delay the onset of heating until 3:30 am. Thereby saving two and a half hours of heating which for a large building represents a significant saving in fuel costs.

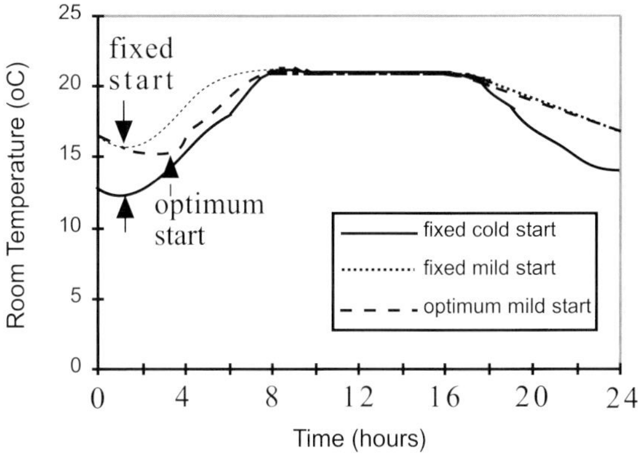

Figure 1.29 Graph of room temperature against time of day showing the benefits of optimum starting

The most difficult parameter to determine for the effective operation of the device is the thermal inertia of the building. Initial estimates of this may need to be modified during the commissioning stage to achieve accurate performance. Some devices monitor their own performance and carry out this adjustment automatically. They are known as self learning optimisers.

Compensated temperature control. Temperature control is achieved in domestic buildings by simply switching on and off the flow of heat to the radiators. When the desired room temperature is achieved the boiler will be switched off and after a short run on period, to dissipate residual heat, so will the pump. This is not practical in large buildings because of the thermal inertia of the large volume of water circulating in the heating system. Swings in temperature about the set point would be too great as the water in the heating system was alternately heated and cooled. Instead, large buildings do not switch on and off the flow of heat to the building but modulate it up or down as the demand for heating goes up or down. This process is achieved using a compensated flow circuit.

Compensated flow circuits vary the boiler flow tem-

JOHNSON CONTROLS
Operations & Maintenance Strategies
Performance Contracting
Fire & Security Solutions
Building Management Systems
M&E Projects
JOHNSON CONTROLS

perature in response to changes in outside air temperature.

Figure 1.30 illustrates the principle. The graph is known as a compensated flow curve. Moving from right to left on the graph, it can be seen that as the outside temperature increases the temperature of the boiler flow is progressively decreased. At 16°C outside air temperature the boiler flow temperature is equal to ambient conditions. In effect the heating is switched off. This temperature is known as the outside air cut off temperature. At this temperature no mechanical heating is required since 6°C worth of heating can be achieved from casual gains in the building from passive solar energy, body heat, lighting and appliances.

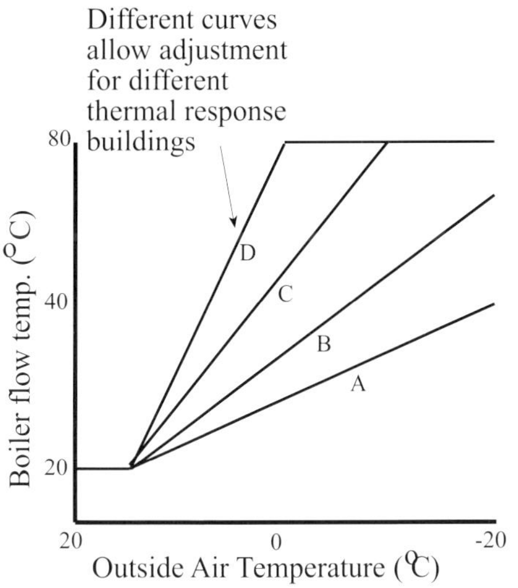

Figure 1.30 Compensation curves

The adjustment in flow temperatures is achieved using a variable temperature (VT) heating circuit. How this is achieved is illustrated in figure 1.31 and IP8. The boiler produces hot water for the constant temperature (CT) circuit at 80°C. Using a three way valve a proportion of this hot water is allowed to pass into the heating circuit. When the demand for heat is high such as on a cold day more hot water will be allowed into the heating circuit. On a mild day less heated water would be allowed into the heating circuit. The radiator temperatures will therefore be hotter on a cold day than a mild one

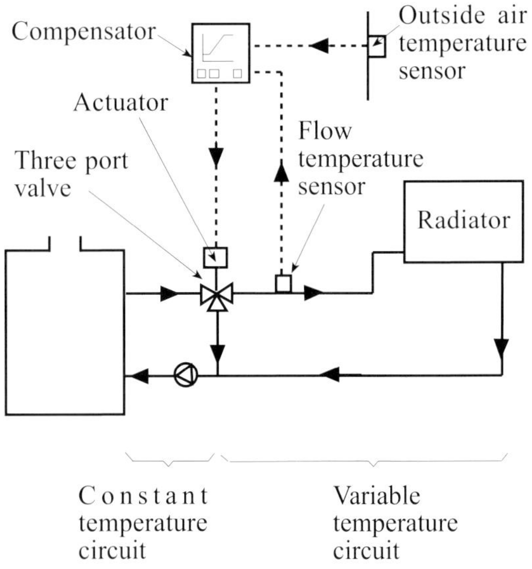

Figure 1.31 Variable temperature flow circuit

ZONING

In large buildings some spaces may need heating and some may not. Differing heating demands within the same building occur for two primary reasons, these are; differences in heat gains and differences in occupancy patterns (hours of use).

For example, south facing rooms will experience solar gains and the heating effect may be sufficient to remove the need for mechanical heating. In this situation the heating to the south side of the building should be shut off. Heating will still be required in north facing rooms. It can be seen then that the building can be split along an east/west axis into two zones one facing north and the other facing south. Further zones can be identified such as those subject to other heat gains. For example, rooms with high occupancy levels or where extensive use of computers is being made.

Individual room by room control of the south side of a building can be achieved using thermostatic radiator valves. An alternative method of zoning, which is more appropriate to larger buildings, is to install separate pumps and flow and return pipe work to supply each zone. A motorised valve, zone temperature sensors and

an appropriate control system are required to control the flow of heat into each zone. As a result there is a capital cost associated with zoning a building. However, these additional costs will be recouped over time in the value of energy savings made. There are also additional benefits to zoning a building such as greater degree of temperature control leading to improved thermal comfort and greater productivity of staff.

Separate heating circuits can also be used to cater for differential occupancy of spaces within a building. One example is in the case of a school which holds night classes. Rather than heat the whole school for this event the school should be zoned and heating can then be supplied to the night school block only.

If the use of parts of buildings are being charged for separately heat meters can be fitted to each heating zone pipe work. These meters monitor how much heat is being taken by the zone from the central boiler plant. Knowledge of this allows accurate costing of out of hours use of spaces to be made

BUILDING ENERGY MANAGEMENT SYSTEMS (BEMS)

All of the control functions discussed previously including boiler step control, optimisation, compensation and zoning can be carried out using a building energy management system (BEMS). A BEMS is a computer based heating, ventilation and air conditioning control system which offers a great deal of flexibility in the way it is set up and operated. It also offers the possibility of close interaction between the operator and the building services systems. The main components of a BEMS are shown in figure 1.32 and are described below.

Outstations are small computers. Unlike dedicated hard wired controllers which only control the functions for which they have been purchased outstations are flexible in what they can do. The outstation can be programmed to perform any or all of the above control functions. The outstation receives information about what is happening to the heating system and the building from sensors. The programme which decides what this information means and what to do as a consequence is called a control strategy. A simple example is if the room temperatures are below the required temperature setting known as the set point. Temperature sensors will signal this to the outstation. Using the logic contained in the control strategy the outstation will send signals to actuators to make the boiler fire and pumps operate to supply heat to the room. The control strategy refers to other rules before carrying this out for example heating will only be supplied if the time is within the occupancy or pre heat period.

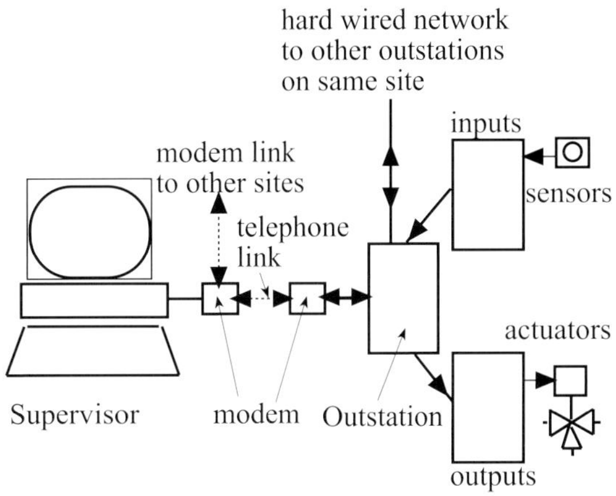

Figure 1.32 Building energy management system components

Sensors are the input devices for the outstation. They are transducers which convert a physical state into an electrical signal. There are two main types; analogue sensors and digital sensors. Analogue sensors return a varying signal to the outstation. For example, a signal in the range 0 to 5 volts from a temperature sensor can be set to represent the temperature range 0 to 25°C.

A digital signal can only take one of two values for example 0 volts or 5 volts. Such a signal can be sent from a switch to represent it being opened or closed. So for example if a boiler was firing a 5V signal would be returned if it had failed then a 0V signal would be returned.

Actuators like sensors can be either analogue or digital. They are devices which turn electrical signals into physical actions using motors or solenoids. An example of an analogue actuator is one which sets the position of a motorised valve. The electrical signal to it may vary between 0 and 5V this corresponds to the fully

shut and fully open positions. Hence a signal of 2.5V would cause the valve to be half open.

A digital signal can be used to open or close a solenoid. For example changing the signal to a pump from 0 to 5V causes a pump to operate returning the signal to 0V causes it to stop.

The supervisor allows human operators to interface with the system. It is a standard personal computer which is loaded with the necessary software to interact with the outstations. The supervisor can be used to programme the outstation with its control strategy. Once this is achieved it is possible to visualise on screen all of the information available to the outstation. So for example room temperatures, the status of boilers, pumps and other equipment such as the position of valves or dampers can all be displayed. This information is displayed graphically so that their interpretation is easily understood with only a small amount of training.

The system constantly upgrades the information it presents and also stores data at the outstation for later inspection. For example, room temperatures over the last 24 hours can be displayed graphically. This is a most useful tool for diagnosing faults and commissioning the heating system following installation. The supervisor is also used to set variables. One example is the inputting of room temperature set points.

Buildings fitted with a BEMS have been found to have low energy consumptions. There are a number of reasons for this. The first is the accuracy of control that can be achieved. The second is the ability of the system to signal heating system faults which may otherwise go undetected causing excessive energy usage. Finally, monitoring and management of energy consumptions is also facilitated by fitting sensors on to the utility meters. This allows logging of energy consumptions which can then be used to prepare reports and note excessive consumptions.

One supervisor can be used to control the operation of many outstations. The supervisor will be located in the office of the energy or building manager. It can communicate with various outstations using the telephone system and network cabling on site. This communication is not limited to outstations and other BEMS systems. It is also possible to communicate with other control systems in buildings. For example it is possible to integrate BEMS systems with security systems. So for example access to spaces using key cards can be monitored. When it is known that all people have left a space the heating can be turned off or turned down to a set back position. Communication can also occur between BEMS and fire systems. Ventilation systems which would cause spread of smoke can be closed by dampers in the event of fire being detected and smoke clearance fans can be turned on.

1.7 Valves

Valves have a role in the commissioning, operation and maintenance of wet indirect heating systems. In commissioning, valves are used to balance the flow of water around the distribution and heat emitter network, in operation, valves are used to direct and control the extent of heat output and finally, in maintenance, valves are used to isolate failed sections of pipework and components for repair. These roles are achieved using the way that valves modify the flow of water in pipes. There are many different types of valve. The following section describes the three valve functions associated with flow modification and gives an example of a specific valve type used to carry out each one. These valves are illustrated in figure 1.33.

- The first function is to stop the flow of water completely. These valves, also referred to as isolating valves, are fitted on both sides of components such as pumps. The valve, when closed, stops the flow of water so that the pump can be removed for repair without having to drain down the entire system.

One type of isolation valve is the globe valve. Water normally flows through a gap in the valve body. Turning a threaded rod by hand or by a motorised actuator causes a plug on the end of the rod to block the gap in the valve body stopping the flow.

- The second function is to regulate the flow of water between full flow and no flow. If the water is heated the amount of heat delivered to a heat emitter can be varied by varying the flow of water. Thermostatic radiator valves (section 1.6) work in this way. One type of flow regulating valve is the butterfly valve.

BELIMO

Simply the best way to drive a valve!

Mixing actuators and motorized ball valves for HVAC water circuits

Linear actuators and lift type globe valves for HVAC water circuits

Benefits of Characterised Control Valves

Belimo ball valve with an innovative characterising disc

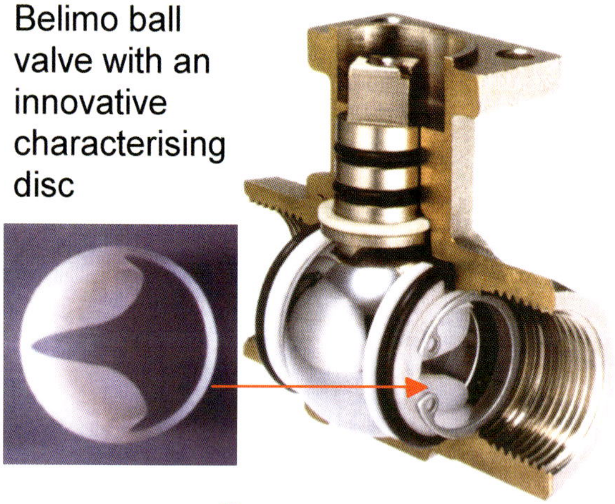

True, equal percentage valve characteristic curve

No initial surge upon opening

Excellent control stabilty

Performance comparable to a globe valve of similar size

Requires fewer pipe reductions

Better partial load behaviour and prevention of the systems tendancy to oscillate giving higher control stability

The Super Compact Control Valve

Combining the very latest in actuator design with innovative valve technology has produced a true full function control device - made possible by Belimo's intelligent linear actuator. The new type of control gives the Super Compact control valve access to the applications of the classic flanged control valve - offering all the same advantages but without the usual bulkiness and weight.

Belimo Automation UK Ltd

The Lion Centre
Hampton Road West
Feltham
Middlesex TW13 6DS

TEL: 020 8755 4411
FAX: 020 8755 4042
E-MAIL: belimo@belimo.co.uk
www.belimo.org

This valve has an internal disk whose diameter is the same size as the bore of the valve. When the disk is positioned across the valve flow is stopped, when the disk is in line with the bore full flow is achieved. The position of the disk is determined by a rod connected to the disk centre pivot and extending out of the valve. Rotation of the rod determines the alignment of the disk and hence rate of flow of water.

- The third function is to direct the flow of water down one of two alternative outlet pipes connected to the valve. For example a three port valve has one inlet and two outlets. Flow entering the inlet can be directed down either of the outlets or shared between them. One example of the use of this is in domestic heating system control. Hot water from the boiler can be diverted either to the heating circuit, dhw cylinder or shared between the two depending on which thermostats are calling for heat.

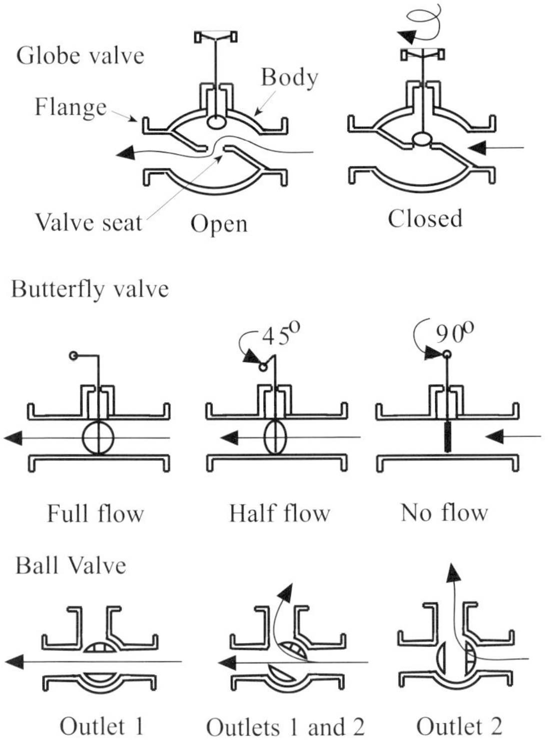

Figure 1.33 Functions and types of valve

An example of a flow diverting valve is the ball valve. The valve body has three ports. Situated at the centre of these in the bore is a ball. This ball has a hole bored through it. Water normally passes through inlet to outlet 1 through the hole in the ball. However, rotation of the ball at first shares flow with outlet 2 then diverts it wholly to outlet 2.

A mixing valve is a three port valve working in the opposite sense. It has two inlets and one outlet. Inlet 1 carries hot water, inlet 2 carries cold water. Variable positioning of the ball will mix the two flows to give a constant temperature at the outlet ranging between the two flow temperatures.

The build quality of valves is an important issue. For example globe valves must seat positively if they are to stop flow. Control valves must be constructed to close tolerances because when they begin to close the pressure increases and there is a tendency for water to seep past the valve. This will result in energy wastage and poor control. This is particularly important when valves carry out more than one of the functions described above. For example globe valves can act as a flow regulation and an isolation valve (page 42).

The setting of valves can be adjusted manually as is the case with most isolation valves. However they can also be set using a motorised actuator which is the case for valves associated with control. The position of the butterfly or ball within the valve is usually indicated by a visual marker outside the valve. This enables maintenance workers visiting the plantroom to assess its current position and check that the position changes in response to a control signal.

1.8 Feed and Expansion

Indirect heating systems using water as the heat transfer medium must have some means of replenishing minor water losses and accommodating the expansion that occurs as water is heated. Domestic systems use a feed and expansion tank. In commercial systems the pipe work is sealed and a pressurisation unit is used to satisfy the requirements for feed and expansion.

Feed and Expansion (FE) Tank is usually sited in the loft so that it is higher than the heating system it serves. Water from the mains fills the tank to a pre set level determined by the ball valve (figure 1.34). If water is

IP8-SCHEMATIC-TWO.ZONE.MULTIPLE.BOILER.SYSTEM

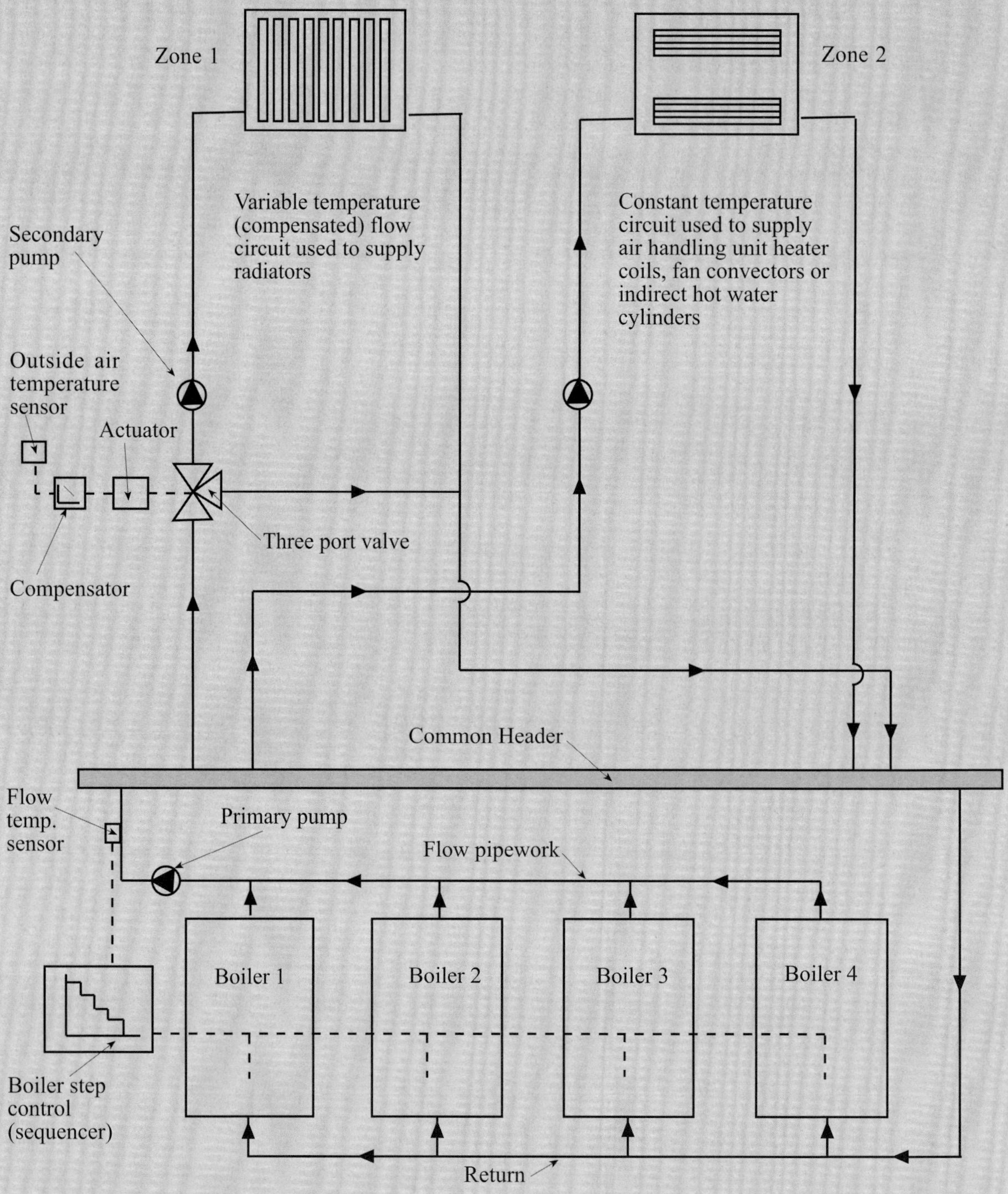

Reproduced with permission from **Hamworthy Heating Ltd.** from their publication 'Energy Efficient Multiple Boiler Systems'

lost from the heating system it is replenished from the FE tank via the feed water pipe. The expansion pipe is connected to the heating system pipe work and terminates in FE tank (figure 1.1, page5). If the heating controls should fail the temperature of the boiler flow could rise causing the water in the system to expand. If the system were sealed the resulting pressure rise would damage the system. However this does not occur since the additional water volume can escape through the expansion pipe. Excess water in the FE tank will be safely discharged outside via the overflow pipe.

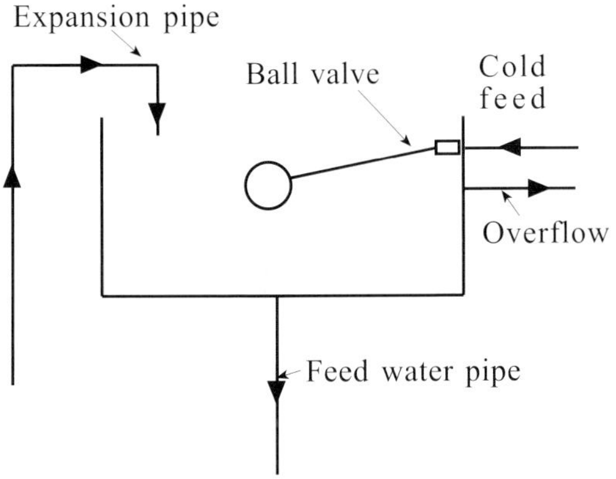

Figure 1.34 Feed and Expansion tank

Pressurisation unit. The pipe work in heating systems incorporating a pressurisation unit has no openings to the outside air. It is sealed. A pressure sensor is fitted into the system as shown in figure 1.35. If the pressure falls too low a pump is operated that feeds mains water into the pipe work. When the boiler fires pressure in the system will increase due to the expansion of water. This increased water volume is accommodated in the expansion tank. The tank is partially filled with nitrogen which is compressed by the incoming water. If, due to a boiler control fault, the pressure increase is too great then an interlock between the pressurisation unit and boilers will switch the boilers off until the cause of the fault can be corrected. Excess pressure is released by allowing water to escape through the pressure release valve. Continuous filling of the system by the pump would signal a leak in the system.

As well as keeping system water levels maintained the pressurisation unit can also pressurise the water in the system. The benefit of this is that the boiler flow temperatures can be increased above 100°C without it boiling. This means that the same volume of water will carry more energy to the heat emitters. In a large building this will result in smaller pipe diameters which are easier to accommodate within service runs. The temperature ranges encountered are;

- Low temperature hot water (LTHW) 70 - 100 °C
- Medium temperature hot water (MTHW) 100 - 120 °C
- High temperature hot water (HTHW) 120 - 150 °C

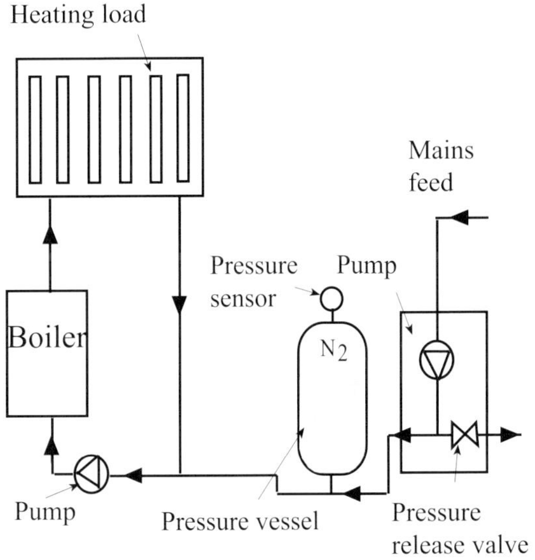

Figure 1.35Pressurisation unit

The high temperatures can be used to provide heat for certain industrial processes such as drying. However care should be taken with positioning heat emitters which will have very hot surfaces and would burn anyone touching them.

2.0 Indirect Warm Air Heating

Air can be used as a heat transfer medium in the same way that water is used to carry heat. The warm air is

NOTES

delivered to the space to be heated in ducts rather than pipes. Air has a much lower heat carrying capacity than water and so ducts need to be bigger in cross sectional area than a water pipe to carry the same amount of heat. The space required for these ducts must be allowed for in the structure of the building, see IP7.

There are three benefits of warm air heating. The first is that ventilation can be supplied to a room along with the heat. The second is that the time taken to warm up rooms to comfort temperatures is less than that taken by a wet system and finally the room warm air outlet terminals take up much less wall space than a radiator. Figure 2.0 shows the main components of a domestic indirect warm air heating system.

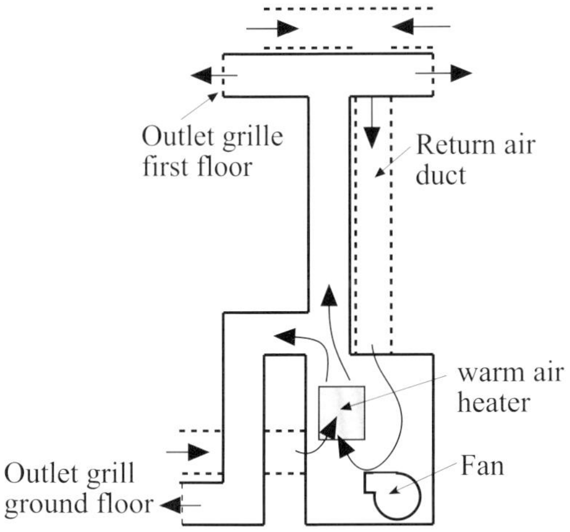

Figure 2.0 Domestic warm air heating system

A centrifugal fan is used to pass air over a surface which is heated by a gas burner. This air is then delivered to the various rooms using ducts. The vertical ducts are sited centrally in the building to achieve economy of duct lengths and ease of accommodating them. Short horizontal runs deliver the air to each room. The outlet grilles should be directed towards external walls to counteract the cold down draughts which occur at windows. Air typically returns to the heater via the stairwell and halls. Gaps are usually created in doors to allow the free passage of return air. Air can also return to the heater via ducting as is shown in the diagram. Return inlet grilles should not be sited in toilets or bathrooms due to the risk of re circulating moisture and odours throughout the building. Fresh air enters via infiltration routes into the building.

COMMERCIAL SYSTEMS

Indirect warm air heating systems for large buildings share many of the characteristics of air conditioning systems which are discussed in detail in section 7.0 Figure 2.1 shows a typical warm air heating system. Outside air is drawn into the air handling unit using a centrifugal fan. It is filtered and heated before leaving the unit to be delivered to each room using ducting. The outlet diffusers may be sited under the window sill to counteract perimeter heat losses. Dampers are fitted to control the flow of warm air out of the unit. The return air outlet is positioned to give a good flow of air across the room, possibly within a suspended ceiling. The return fan is smaller than the supply fan so that the rooms become slightly pressurised. This helps to prevent the ingress of draughts into the building.

An advantage of warm air systems is that in summer the heater coil can be turned off and free cooling obtained by bringing in fresh outside air. By pass ducting will be required to avoid any heat recovery devices to ensure that the incoming cool air is not preheated by the outgoing stale air.

If the exhaust air quality is good then the majority of it will be re circulated back into the inlet. The remainder of the air supply will be fresh air. In this way some of the energy used to heat the air will be recovered. If the air quality is poor then 100% fresh air will be used. To avoid wasting energy heat should be recovered and used to pre heat the supply air (section 4.4).

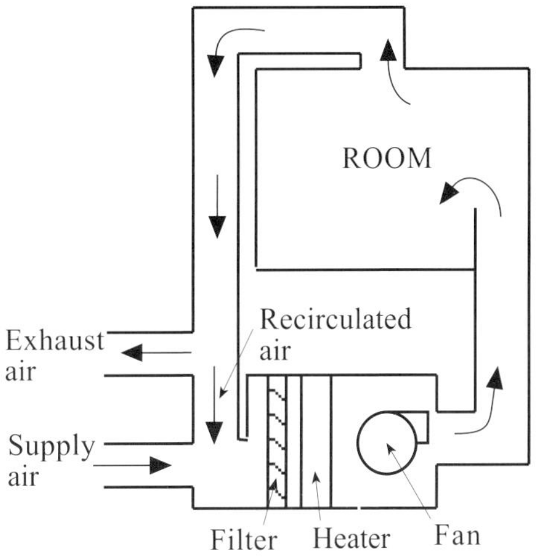

Figure 2.1 Commercial warm air heating system

IP9 - HEAT.TRANSFER.MECHANISMS

Heat naturally flows from a body at a high temperature to bodies at a lower temperature. It is as if the universe is trying to balance out the temperature of all the objects contained within it. In the far future it can be predicted that everything will have the same temperature and heat transfer will stop. Until that day we can use heat transfer mechanisms to make our heating systems work.Heat moves by three mechanisms; conduction, convection and radiation.

All of the universe would really like to be one temperature. To achieve this, heat flows from high to low temperature objects.....Until this final temperature is met we can use the effect to make our heating systems work!

CONDUCTION

This is the transfer of heat through solid materials. This mode of heat transfer is often descibed by imagining that the atoms of a material vibrate. The hotter the material the greater are these vibrations. Conduction is explained by the transfer of these vibrations from one atom to the next, moving from the warm end of the material to the colder end. An example of conduction in buildings is the loss of heat through the building fabric from the warm interior to the cooler outside. Conduction of heat is also used in calorifiers and plate heat exchangers. Close contact is required between the hot and cold fluids to ensure good heat transfer is made.

CONVECTION

This is the transfer of heat within liquids. Here we can consider both air and water to be a type of liquid. Using the concept of vibrating atoms introduced above. It follows that the atoms in warm air will be more widely spaced due to their increased vibrational amplitude than those in cool air. As a result the warm air will be less dense, and so will float in the cooler air causing it to rise. When the warm air becomes cooled once more it will have the same density as the rest of the room air and convection will stop. Convection currents carry heat away from radiators and convector heaters. Convection also causes stratification of temperatures in tall spaces (see page 51). Downward convection currents occur when cold air is present in a space. This is a particular problem along the perimeter of highly glazed buildings. Downward cold convection currents from the windows can produce uncomfortable draughts. Which must be countered by a perimeter heating system (see page 23).

RADIATION

Any warm object will radiate heat to another object at a temperature lower than itself. This heat is in the form if infra red radiation which does not require gaseous or solid material for its transfer. When an object absorbs infra red radiation its temperature will be increased. Infra red radiation is the warmth we feel from the sun. It is also used as the heat transfer mechanism by radiant heaters. Thermal radiation is a similar form of energy to light. As a consequence just as light can be guided and directed using reflectors so can thermal radiation.

MASS TRANSFER

An additional mode of heat transfer used in buildings is that of mass transfer. In this a fluid such as air water or steam is used as a heat transfer medium. By moving the air or water from one place to another the energy held by it is also transferred. Following the mass transfer the occupants in the room are heated by one of the above three mechanisms.

3.0 Direct Heating Systems

Direct heaters, which were introduced in section 1.0 give out their heat by a combination of convection and radiation directly into the space they are heating (see IP9). Common domestic gas fires are one type of direct room heater which aim to encourage both forms of heat output. However, for purposes of discussion in this book direct heaters will be grouped in to two main categories; those which give out most of their heat by radiation and those which give out most of their heat by convection.

This section will consider common types of domestic and industrial convector and radiant heaters and will conclude by looking at direct water heaters

3.1 Convector Heaters

Convector heaters give out their heat using the natural buoyancy of warm air. Room air comes into contact with a hot surface. When warmed this air becomes less dense and so rises out of the heater to warm the room. Some devices use fans to increase airflow across the heated surface. This increases heat output and reduces the time taken to heat the room.

DOMESTIC HEATERS

Gas Convector Heaters. These heaters are room sealed combustion devices. The combustion air is taken from outside the building via a balanced flue rather than from the room itself (figure 3.1). As a consequence heaters are usually installed on outside walls. Heaters can be installed on internal walls using longer lengths of flue and exhausting the flue gasses using a fan. To avoid the complexity associated with electrical wiring, basic models are fitted with a piezoelectric spark ignition. Temperature control is achieved manually using a variable burner setting. The only connections required by the heater therefore is a gas supply. To improve the level of control some models incorporate a room thermostat and a time switch the operation of which requires connection to an electrical supply.

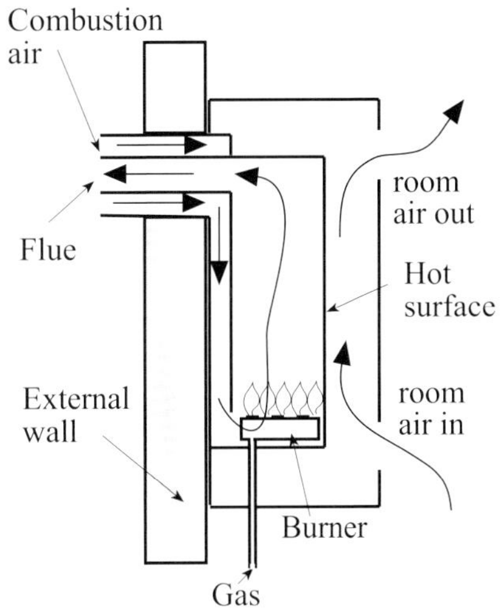

Figure 3.1 Gas convector heater

Electric storage heaters. This type of heater comprises of a metal casing which is filled with dense blocks (figure 3.2). The blocks have a high specific heat capacity (see IP6). This means they can store large amounts of heat for each degree of temperature through which they are heated. The blocks are heated by electrical heating elements which run through them. Charging with heat usually takes place overnight to take advantage of cheaper night-time electricity tariffs. The amount of heat given out by the storage heater in the day de-

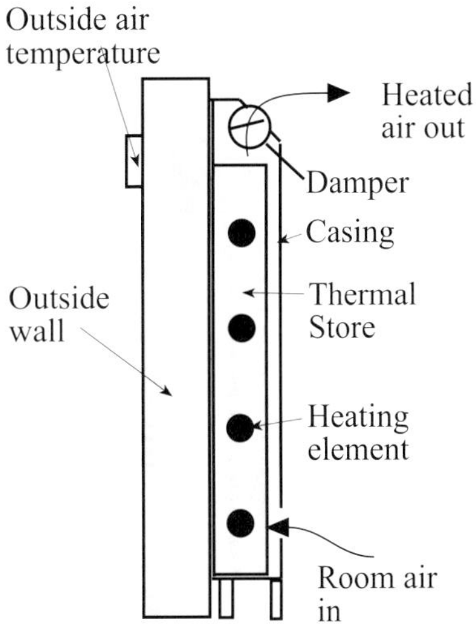

Figure 3.2 Electric Storage Heater

NOTES

pends on how much heat is stored by it overnight this, in turn, depends on the time over which the heating elements are operated. Charging time can be set manually using a dial on the heater graduated in hours or it can be carried out automatically. Automatic control requires a controller which monitors outside air temperatures. If the air temperature is low, indicating that more heating will be required the next day, then the heating current will be allowed to flow for longer.

Heat output is by convection currents passing across the casing and through the heater. If the room becomes sufficiently warm further heat output can be restricted by closing a damper to block the internal convection path. Fanned storage heaters are available. These utilise well insulated casings to minimise heat losses. Heat output is via forced circulation of air through the unit. These storage heaters give a better degree of control but utilise additional electricity to operate the fan.

COMMERCIAL WARM AIR HEATERS

Direct heaters used in commercial or industrial buildings are bigger in scale than domestic heaters so that they can satisfy the increased heating demand of the large spaces that are used by industry. They are usually less decorative but are more robust. They are often mounted at high level to free up floor space which is more economically used as production area.

Floor standing cabinet heaters. Figure 3.3 shows a diagram of a floor standing cabinet heater. Air is drawn from the space by a fan and is passed over a surface heated by a gas burner. The hot air is then directed back into the room via cowls. Adjustable vanes within the cowls allow further variations to be made to the direction of the warm air jet. Combustion air for the burner is drawn into the heater from the adjacent space or it may be taken from outside using ducting. The times within which the heater can operate is controlled using an optimiser as described in section 1.6 (page 35). The space temperature is controlled using a thermostat which monitors the temperature of the air entering the heater. If the temperature is above set point the burner is switched off or modulated down.

It is not advisable to use warm air heaters when the space to be heated is draughty. This is because comfort is achieved by contact with the warm air. If this warm air is regularly removed from the space for example by opening loading bay doors then comfort will not be achieved and energy will be wasted. Care should also be taken where the space to be heated is tall. This is because warm air rises and hence stratification will take place.

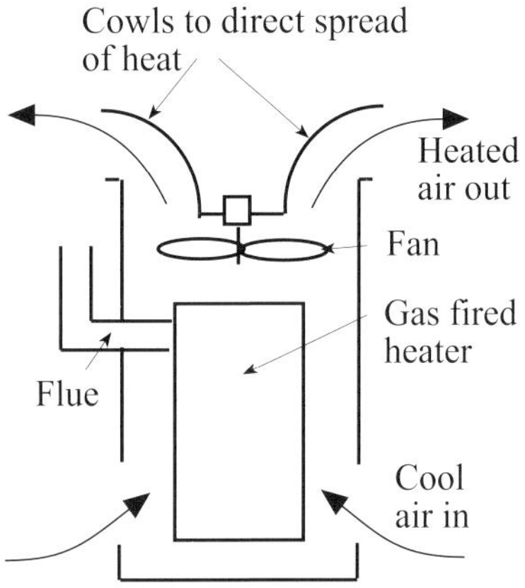

Figure 3.3 Cabinet heater

Stratification is the creation of a temperature gradient which increases between the floor and roof. The temperature adjacent to the roof in a 4m high space may be as high as 30°C. At low level, in the zone occupied by people, the temperature is 21°C. The high air temperature next to the roof is unnecessary for comfort and will increase heat losses through the roof.

The problem of stratification can be avoided using de stratification fans (figure3.4).

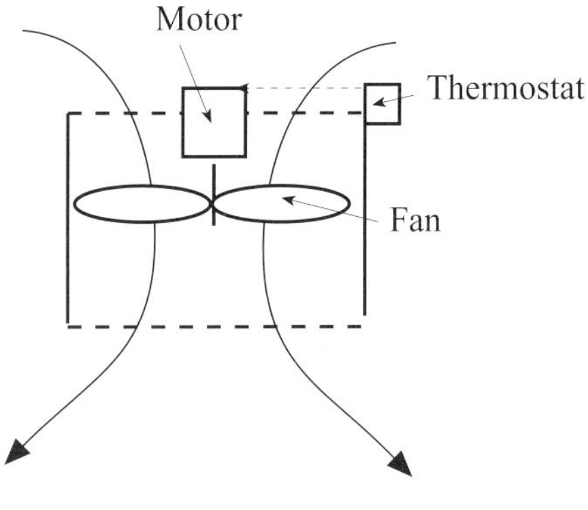

Figure 3.4 De-stratification fan

NOTES

These are fans mounted at high level in the space. An inbuilt thermostat monitors the air temperature next to the fan. If the temperature rises above a threshold temperature setting such as 27°C then the fan operates pushing the warm air back down into the occupied space.

Unit heaters. Unit heaters are a smaller version of cabinet heaters which are mounted at high level in a space (figure 3.5). Air flows through the heater, driven by a fan. It is heated as it passes through a series of venturi shaped plates made hot by a gas burner. Adjustable vanes on the warm air outlet directs the heat down into the occupied space. One of the benefits of this type of heater is that they do not occupy any floor space. Another benefit is that they help to avoid stratification by taking in air at high level and directing it down towards the occupied space.

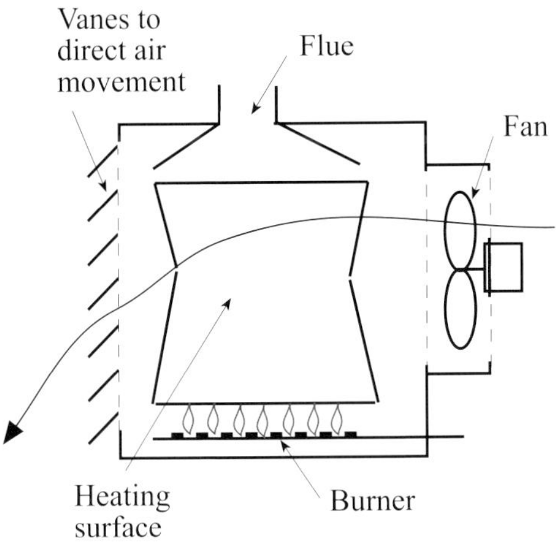

Figure 3.5 Unit heater

Heating and Ventilation. The warm air heaters described above use recirculated room air as the medium for heating. However, fresh air can be supplied to the heater using a length of ducting so that heating and ventilation can be provided to the space at the same time.

Figure 3.6 shows, using a roof mounted unit heater as an example, that three modes of operation are possible by varying the position of dampers in the heater.

- Recirculation. If the room air quality is good, 100% recirculated air can be used. The high level input would de-stratify the space. If ventilation is required during occupancy 100% recirculation can still be used during the pre occupancy period to shorten the building warm up time.

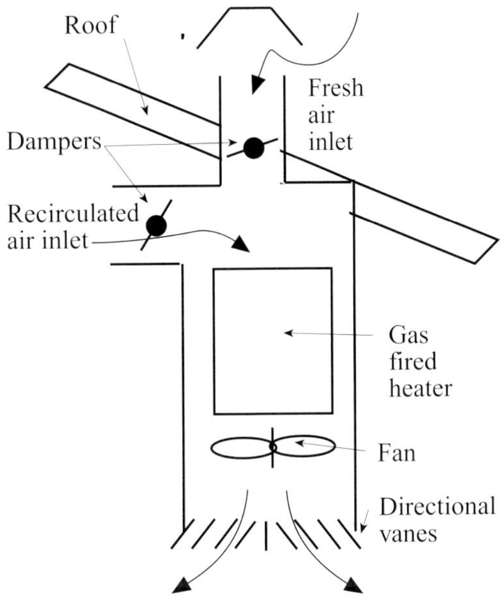

Figure 3.6 Ventilation unit heater

- Ventilation. If the room air quality deteriorates then a mix of fresh and recirculated air can be used. The proportions in this mix can be determined using room air quality sensors fitted into the inlet duct of the heater. Poor air quality would result in greater proportions of fresh air being used.

- Full fresh air. If 100% fresh air is to be used for extended periods it would be economical to add an extract duct to the heater to enable the incoming and outgoing airstreams to pass through an air to air plate heat exchanger (see section 4.4). This would save energy by pre-heating the incoming airstream using energy contained in the exhaust airstream. In summer the use of 100% fresh air can also provide free cooling by switching off the heater and bringing in outside air if it is cooler than the space temperature.

3.2 Radiant Heaters

All hot surfaces radiate heat to objects cooler than

NOTES

themselves (see IP9). When a person is near a heated surface this radiant heat is felt as warmth. The essential element of a radiant heater is, therefore, an exposed heated surface. The human body is very sensitive to radiant heat and feelings of warmth are readily experienced in its presence (see IP5). As a result it is possible to feel comfortably warm in draughty spaces such as warehouses or workshops if a source of radiant heat is present, even if the air temperature is low.

Radiant heaters have a low thermal capacity. This means they will heat up quickly giving a quick response time, creating comfort shortly after being switched on. The heating effect is principally by radiation but eventually the radiant heat absorbed by people and objects will be re-emitted by convection. This will result in a gradual increase in room air temperatures

Radiant heat is a form of electromagnetic radiation. This is the same form of energy as light but at a different wavelength. Because of this it behaves like light. It can be reflected to where it is needed but also blocked by objects in its path. This latter point means that people can be shadowed from its warming effects. One example is the shading of customers by the tall shelves in a DIY store. To solve this problem radiant heaters are usually sited over the aisles where customers circulate. Since the heaters do not rely on warm air for their heating effect standard thermostats which detect room air temperature cannot be used to control them. Control is achieved using a black bulb thermostat. This is an electronic or bimetallic strip thermostat built into a black plastic hemisphere. The hemisphere absorbs the radiant heat which in turn warms the air trapped inside the globe. It is this enclosed warm air that the thermostat senses to provide control over the heaters.

There are two main types of direct radiant heater these are plaque heaters and radiant tube heaters.

Plaque heaters. Are comprised of a flat surface (figure 3.7) heated either by an electrical element or a gas burner. The surface of the heater is warmed until it radiates heat. Some units become red hot. This high surface temperature means that they must be fixed at high level, usually above three metres in height, so that the occupants of the building cannot be burned. Lower temperature plaque heaters are available which have much lower mounting heights. Plaque heaters are most often used to give spot heating to a particular location. This could, for example, be a workstation sited in the middle of a warehouse. Comfort can be achieved without heating the entire space.

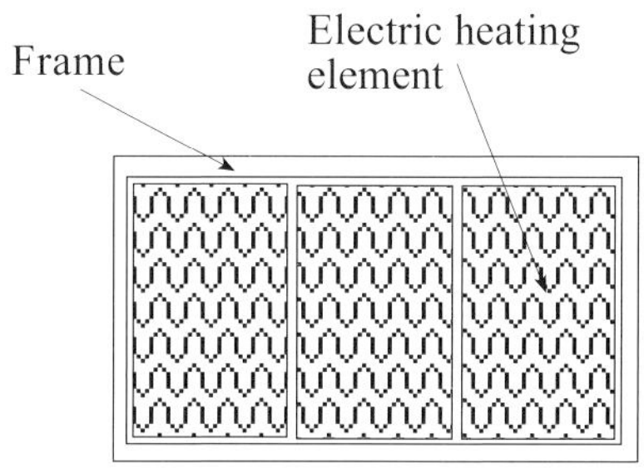

Figure 3.7 Radiant plaque heater

Radiant tube heaters are composed of a gas burner connected to a steel tube (figure 3.8). The burner directs the flames it produces along the inside of the tube. As a result the tube becomes hot and starts to emit radiant heat. Sometimes the tube is formed into a U shape doubling the surface area for heat output. A reflector above the tube directs the heat downwards. Radiant tube heaters, as with all heat emitters, have a limited area of influence. To heat a large space evenly an array of radiant tube heaters must be arranged in a similar way to luminaires used to light a space.

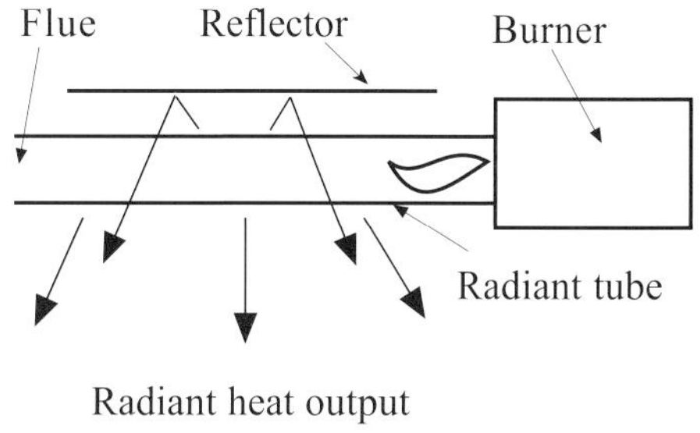

Figure 3.8 Radiant tube heater

Energy issues. Radiant heaters can result in significant savings in energy when used appropriately. The three situations when this occurs are;

Tall buildings - where excessive stratification would occur with warm air heating.

Spot heating - For example heating a manned workstation in the middle of a warehouse. Comfort can

NOTES

be achieved without heating the entire space.

Draughty buildings - High air change rates make warm air heating impractical and wasteful.

3.3 Direct Water Heating

The heaters in a direct heating system are dedicated to providing space heating. Because of this some form of direct water heater is also required. In addition, it is more economical and energy efficient to separate water heating from the indirect space heating system by providing hot water from a direct fired water heater. This avoids the low efficiency of indirect heating systems when firing at low loads when hot water only is required.

Direct water heating in domestic buildings takes one of two principle forms these are storage systems and instantaneous systems.

Storage systems. An example of a storage system is the use of an electric immersion heater to heat the water in a dhw cylinder. The immersion heater is a sealed element through which electricity is passed causing it to heat up. The immersion heater is fitted into the storage cylinder to make direct contact with the stored water. A typical domestic immersion heater would have a power rating of 3 kW.

The immersion heater has an in built thermostat to control the domestic hot water temperature. This switches the current on or off, as required, to maintain the pre set temperature, typically 60-65°C. The duration of operation of the heater can be controlled manually using a switch or automatically using a time switch.

Instantaneous water heaters warm the water as it is drawn from the tap so that it only heats the amount of water that is required. Gas fired water heaters are constructed like small boilers with a burner and heat exchanger. The water input to the heat exchanger is from the mains. Output is to the taps. Turning on the hot tap causes a pressure drop in the pipe work which, after a short delay, signals the burner to fire. Electrically heated units are also available. Electrical heaters like the gas heaters can supply all the taps in a small building. Small units can supply individual sinks, their small physical size means they can be fitted into the hot water pipe work under the sink. Their use in this way at the point of use avoids heat losses that occur from long pipe runs.

COMMERCIAL SYSTEMS

In large buildings the distance between the source of hot water and the taps is large. This can result in excessive heat losses from the pipe work as the water runs from the source of heat to the taps. One way of avoiding this is to decentralise the hot water system and to provide sources of hot water throughout the site corresponding to demand. Small storage or instantaneous heaters can be used as described above. For larger demands direct fired water heaters should be used. These devices are cylindrical in shape containing a small volume of stored water (figure 3.9). The gas burner is sited at the bottom of the unit. The heat exchanger and flue run up through the centre of the stored water volume and so excellent thermal contact is made between the source of heat and the water. Any heat losses from the flue also pass into the water. The exterior of the tank is well insulated to avoid standing heat losses. The volume of water stored is large enough to satisfy peak demands but small enough to reduce to acceptable levels, the standing losses that occur overnight as the stored volume of water cools down. As a consequence of this and the other features described above direct fired water heaters operate at an efficiency of approximately 90%.

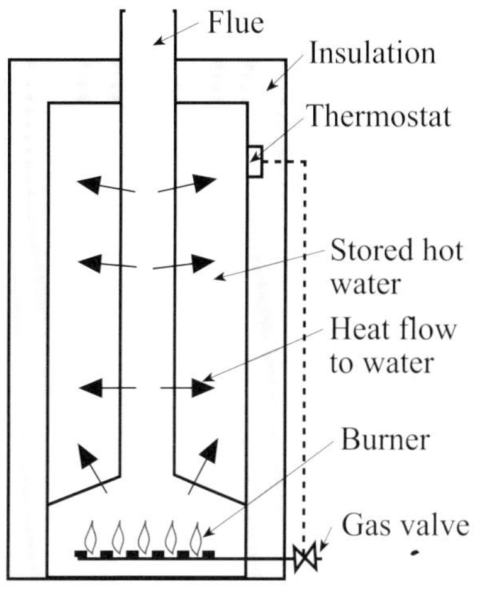

Figure 3.9 Direct fired gas water heater

NOTES

V E N T I L A T I O N

4.0 Introduction

Ventilation is the replacement of stale air in a building with fresh air. Ventilation is a vital requirement for the comfort and health of building occupants (see IP10). The reasons why ventilation is so important can be listed as;

- to supply oxygen for breathing (respiration)
- to dilute pollutants such as body odours and exhaled carbon dioxide.
- to remove unwanted heat from the building.
- to supply air for combustion appliances.
- to lower the relative humidity and so avoid condensation.
- to clear smoke in the event of fire.

There are two ways of stating the quantity ventilation supplied to a space. The first relates to the occupants of a building and is specified as the number of litres of fresh air that is delivered to the room per second per person (l/s/p). Each of the functions of ventilation listed above require different rates of air supply. For example the need for respiration can be satisfied by supplying air at a rate of 0.2 l/s/p. Dilution of body odours from sedentary occupants requires 8 l/s/p. Dilution of odours such as tobacco smoke requires much higher ventilation rates. For very heavy smokers 32 l/s/p is required. For estimation of typical ventilation requirements the value of 8 l/s/p is used.

The second method of quantifying ventilation is to use the air change rate. The air change rate is the number of times the air in a room is completely changed by fresh air every hour. The units are therefore air changes per hour (ac/h). The recommended air change rate in a typical mechanically ventilated office is in the range 4 to 6 ac/h. For dining halls and restaurants the recommended air change rate is between 10 and 15 ac/h. The CIBSE Guides give recommended air change rates for a range of situations.

4.1 Domestic Ventilation

Ventilation of domestic buildings is mostly provided by infiltration. Infiltration is the movement of air in and out of the building via cracks and gaps in the building envelope. Air movement is driven by natural forces such as wind pressure and differences between inside and outside temperatures. Unfortunately infiltration is uncontrollable and, in windy conditions, can lead to excessive ventilation rates resulting in draughts and high ventilation heat losses. Conversely it can also lead to inadequate ventilation on still, warm days. To avoid these problems a four stranded strategy is recommended :-

- The building envelope should be built as airtight as possible to reduce uncontrolled infiltration (see p62). This can be achieved by sealing known infiltration routes such as the gaps between doors/window frames and the masonry, the jambs of door and window openings, around service entries and loft hatches.
- Purpose built openings should be provided to supply background ventilation. Trickle slot or tube ventilators which provide an air movement path through the wall or window heads are suitable. These devices should be fitted with a manual damper to adjust the size of the ventilation opening. Some trickle ventilators are fitted with dampers which close automatically if the airflow rate through the ventilator becomes excessive which may occur on rainy days.
- Windows should be openable to provide short term rapid ventilation of rooms.
- The building regulations section F1 requires that extract fans are used in rooms where moisture and odours are created such as bathrooms, kitchen and toilet. The fan should be sited close to the source of

IP10 - INDOOR.AIR.QUALITY

The occupants of a building are protected against the extremes of climate, noise pollution and to an extent external air pollution by the building fabric. However the internal environment itself acts as a collector and concentrator of a range of pollutants which are known to be harmful to health. Since the majority of people spend over 90% of their lives indoors the quality of indoor air is of great importance for a healthy life. The range of problems that are encountered range from discomfort to death. The former being due to the presence of odours and the latter due to the presence of airborne carcinogens. Cigarette smoke is a source of both categories. Table IP10 lists some of the pollutants found in indoor air.

Category	Example/Quantification
Fibres	Asbestos fibres, no longer used but materials using asbestos still exist in older buildings. Inhalation causes lung disease and cancer. Quantified as number of fibres per litre of air (f/l).
Gases	Carbon Monoxide, produced by incomplete combustion of gas in heating and hot water appliances. In sufficient quantities causes asphyxiation and death. Quantified as a fraction of air volume using parts per million (ppm).
Radiation	Radon gas, a radioactive gas arising from uranium in the ground. Inhalation of radon decay products causes lung cancer. The problem is greatest in houses built over granite. Quantified using radioactivity (Bequerels)per cubic metre of air (Bq/m^3).
VOC's	Formaldehyde. A vapour given off by the binding agents in chipped wood products. Inhalation causes irritation and may lead to allergic reactions such as asthma. Quantified using ppm.
Pathogens	Legionella bacteria. Grows in poorly maintained water based air conditioning and hot water services. Inhalation causes a flue like infection which can lead to pneumonia and death.

Table IP10 Examples of indoor air pollutants

THE SOURCES OF POLLUTION

The sources of indoor air pollution are;

- The materials used to construct the building e.g. volatile organic compounds (VOC's) used as glues to bind materials together such as chipboard
- The building occupants themselves e.g. body odours and carbon dioxide
- Processes being carried out in the building e.g. ozone, pigments and solvents emitted from photocopiers, faxes and computer printers
- External sources e.g. traffic pollution from nearby roads

Humans spend 90% of their lives indoors. The quality of indoor air is therefore of great importance for comfort and health

SOLVING THE PROBLEM

- Avoid using materials or processes which give off pollutants
- Use extract ventilation to remove pollutants at source
- Use supply ventilation to dilute the quantities of pollutants down to acceptable levels
- Use filtration methods to remove pollutants from the air (see section 7.1)
- Seal the envelope to avoid unwanted ingress of polluted air

moisture or pollution and operate to give the specified extraction rate. For example these rates are 15 and 30 litres per second for bathrooms and kitchens respectively. The extract fans should be fitted with appropriate controls to limit the time it runs. One example of control is to fit a humidistat to the bathroom extract fan. When the relative humidity rises over a set threshold the fan will operate. When the humid air is cleared then the relative humidity will fall and the fan will switch off. This will stop heat loss caused by the unnecessary extraction of warm dry air.

Most domestic extract fans are simple propeller fans fitted through the window or wall, possibly incorporating a short length of ducting.

Heat recovery extract fans are a method of extracting stale air and supplying fresh air to single rooms without wasting all the heat contained in the exhaust air. The unit incorporates two fans one to extract air through the unit and the other to bring fresh air into the building in the opposite direction (figure 4.1).

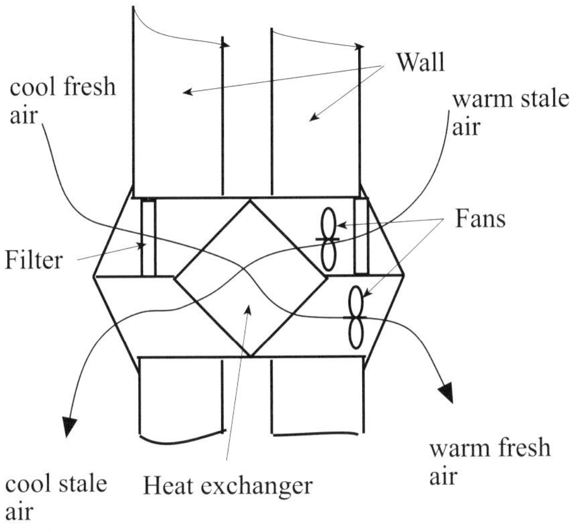

Figure 4.1 Ventilation unit with heat recovery

Both airflows pass through an air to air plate heat exchanger. The two airflows are kept separate and only come into thermal contact with each other. Heat from the outgoing warm stale air passes through thin separating plates in the device and warms the incoming cool fresh air. More details are given in section 4.4.

Whole house heat recovery. The single room unit described above can be extended to provide ventilation and heat recovery to all the rooms of the house (figure 4.2). A fanned heat exchange unit is sited in the loft or within the cupboards above the cooker extract hood. Warm stale air is extracted from all the rooms of the house and passed outside through the heat exchange unit. Incoming air is passed through this unit and is therefore pre-heated before being discharged into each room using ducting.

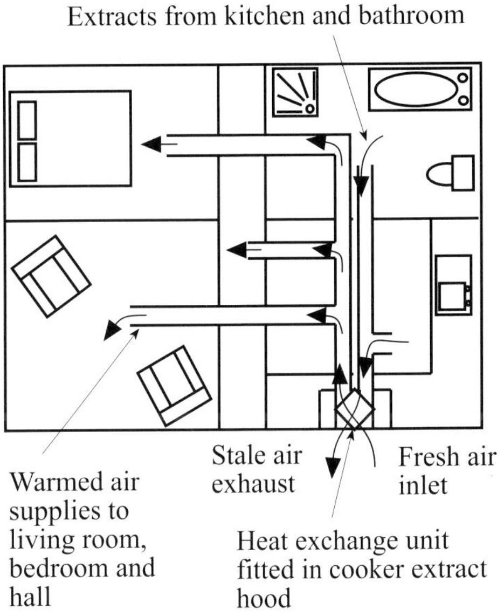

Figure 4.2 Whole house ventilation with heat recovery

4.2 Ventilation of Commercial Buildings

Ventilation of small commercial buildings is carried out in the same way as that for domestic buildings. However as the building size and complexity increases the demand for mechanical ventilation becomes more necessary. A number of situations can be identified where mechanical ventilation is essential these are;

- Densely populated rooms, where the space per occupant is less than 3.5 m^3 per person
- Deep plan buildings. Natural cross ventilation

5

Keeping tabs on Energy Efficiency

Airtightness of buildings

The motto for ventilation and air leakage is 'build tight – ventilate right'. Adequate ventilation is very important, but too much uncontrolled air infiltration causes discomfort and energy waste.

Airtightness can be measured by pressurising buildings and measuring how much they leak. Ideally, most buildings should leak less than 7.5 m^3/h for each m^2 of envelope area when pressurised to 50 Pascals (Pa). Testing shows that a figure of twice this value is typical, and some buildings are four times as leaky.

Airtight buildings improve occupant comfort, save energy and reduce risks of structural damage. As buildings become better insulated, an increasing proportion of the heat loss is through infiltration.

Benefits of airtight buildings

Occupant benefits

- Occupants will not have any discomfort caused by draughts.
- In very leaky buildings internal temperatures might not even reach comfort levels during the winter.
- It is likely that the productivity of the work force will improve since staff feeling comfortable in their working environment will get on with their work, while those who are not will continue reacting to their discomfort.
- With a well-designed mechanical ventilation system contaminants from the outdoor air will be filtered etc.

Energy and environmental benefits

- Energy costs and the resulting carbon emissions will be much higher than necessary in buildings with high air infiltration rates.
- The reduction of air infiltration in a building will allow designers of HVAC systems to reduce the size of the plant and even the size of the plant rooms.

Further information

For further information on building-related energy efficiency, all it takes is one free phone call.
Take action.
Call Action Energy today.
0800 58 57 94
www.actionenergy.org.uk

is difficult to achieve through buildings deeper than 15m from outside wall to outside wall. Single sided natural ventilation is also ineffective beyond 6m from an outside wall.

- Where windows cannot be opened for example due to external traffic noise or pollution outside.

- Rooms where accurate control of temperature or humidity is required.

- Rooms where pollution is created at a faster rate than can be cleared by natural ventilation.

- Where pressurisation of rooms is required e.g. in clean rooms to keep dust laden draughts out.

- Tall buildings which are sealed because wind and stack driven natural ventilation would be excessive

VENTILATION SYSTEMS

There are three methods by which buildings can be mechanically ventilated. These are:

Extract ventilation. This method of mechanical ventilation is used to remove pollutants such as moisture, odours and heat from occupied spaces. A propeller fan is used, mounted in a window pane or through the wall in a short length of ducting, to drive air out of the building. The fan must be sited near the source of pollution so that it can be removed directly from the building without it crossing the occupied space. Whenever air is mechanically removed from a building fresh air will enter to replace it. This make up air will normally enter through cracks and gaps in the building fabric. Alternatively purpose built perforations can be made. In this case the incoming air can be heated using a heater across the path of the incoming airflow. Controls should be fitted to the extract fan to limit its duration of operation to those times when it is needed.

Specialist extract systems for kitchens, laboratories and other industries are available which use fume collection hoods or cabinets. These are connected to outside using ducting and therefore require the use of an axial flow or centrifugal fan.

Supply ventilation works in the opposite sense to extract ventilation in that a fan is used to drive air into the building. The air supplied dilutes pollutants but it cannot directly remove them from the building. The fresh/stale air mixture leaves the building via cracks and gaps in the building envelope.

Supply systems deliver fresh air to the building at high speeds. This is acceptable in summer when high air speeds have a cooling effect but in winter the rapid movement of cold outside air into the building would cause discomfort. This discomfort can be avoided by passing the incoming air over a heating coil before it enters the occupied space. Supply ventilation pressurises the space it is serving, this is useful where entry of draughts and odours from outside needs to be prevented.

Balanced ventilation is a combination of both supply and extract ventilation. Each space is served by supply and extract ducts. The air movement through which is fanned. This system allows a consistent supply of fresh air to be supplied to each room using a configuration similar to the indirect warm air heating systems discussed in section 2.0 and air conditioning systems described in section 7.0.

By making the supply fan more powerful than the extract fan the space will be pressurised avoiding ingress of dust laden draughts. The most important advantage of balanced ventilation systems is the ability to carry out heat recovery between the exhaust and supply air streams.

Air to air heat recovery systems are described in section 4.4.

4.3 Fans

Air movement in warm air heating, ventilation and air conditioning systems is made possible by the use of fans. A fan creates air movement using a rotating vane driven by an electric motor. The casing in which the impeller rotates also has an effect on the air movement characteristics. There are two basic configurations of fan available each with different operating characteristics.

6

Keeping tabs on Energy Efficiency

Energy efficiency in mechanical ventilation

The amount of energy consumed by mechanical ventilation systems can be substantial but is often overlooked – in offices, fans and pumps typically use as much electricity as lighting.

Key design issues

- Do not use unnecessarily high air velocities in ductwork. High velocities result in high fan pressure requirements and thus high powered fans.
- Do not supply more air than is needed – or at times when it is not needed.
- Select an efficient fan.

Early design decisions on space allocation for ducts can be critical, in order to minimise unnecessary bends and junctions which increase resistance to air flow.

Lower air velocities mean quieter systems with larger ducts. Typically, choosing a low velocity rather than medium velocity will reduce running costs by 70%. The additional ducting cost will be recovered in about five years.

Supplying the correct amount of air is partly a question of design air volumes and partly one of providing appropriate controls.

A useful way of checking the efficiency of a mechanical ventilation system is to calculate the specific fan power. This is the power of the fan in kW divided by the flow rate in m^3/s.

System energy efficiency	Specific fan power
High efficiency	<1.5
Medium efficiency	1.5 to 4
Low efficiency	>4

Further information

For further information on building-related energy efficiency, all it takes is one free phone call.
Take action.
Call Action Energy today.
0800 58 57 94
www.actionenergy.org.uk

Axial flow fans have the impeller connected directly to the drive shaft of the motor (figure 4.3). This means that the airflow passes over the motor and is parallel to the axis of the fan.

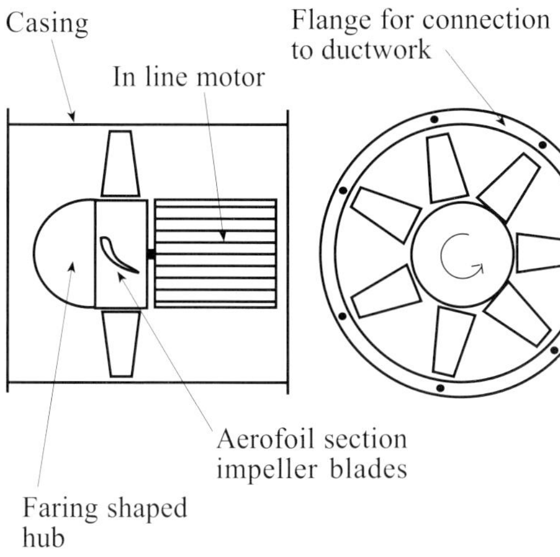

Figure 4.3 Axial flow fan

The impeller blades can be one of two types. The first is a simple propeller where the blades have a uniform cross section throughout their length but are twisted so that when the impeller rotates the air that it comes into contact with is pushed from the leading to trailing edge of the blade. The momentum built up carries the air out of the fan. The volume of air moved depends on the speed of rotation of the fan and the number of blades. Rotational speed is however kept below 30m/s blade tip velocity as the noise generated by the fan becomes unacceptable. Propeller fans do not generate a large pressure difference and so they cannot move air through ducting longer than approximately 45-55 cm. However they are effective at moving air through free openings such as window extract units or wall extract units incorporating short lengths of ducting.

The second blade arrangement is more complex having an aerofoil cross section and twisting from one end to the other. In the same manner as an aircraft wing, the aerofoil bladed impeller generates increased air movement over its upper surface. This creates a greater pressure difference from one side of the fan to the other and so the fan can move air along a system of ducting. The efficiency with which the aerofoil bladed fan converts electricity into air movement is higher than the propeller fan but can be further increased using stationary radial guide vanes across the inlet or exit of the fan. This reduces swirl and so gives a more even flow of air.

Centrifugal fans have a completely different impeller and casing arrangement to axial fans. The impeller blades rather than being perpendicular to the axis of rotation are parallel to it and are arranged into a drum like configuration (figure 4.4). Air is drawn into the fan parallel to the axis of the impeller. The rotation of the impeller causes the air to leave the fan at right angles to the direction of entry. The air being driven by centrifugal force and collected by the volute casing. Changing the arrangement of blades within the impeller changes the characteristics of the fan. For slow, low pressure applications such as is required in noise sensitive environments, the leading edge of the blades faces backwards, away from the direction of rotation. For maximum pressure generation such as is required for moving air through very long lengths of ducting, the leading edge is forward facing.

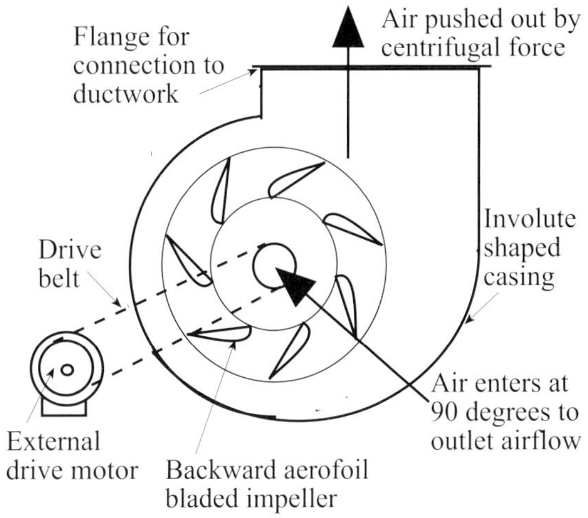

Figure 4.4 Centrifugal fan

4.4 Heat Recovery

Mechanical ventilation gives good control of ventilation rates and hence air quality. However, the air removed from the building carries with it the energy used to warm it up to room temperature. It therefore repre-

IP11 - THE.FAN.LAWS

Three factors of fan performance are of interest to us, these are;

What is the volume of air moved by the fan? (determines the volume of air the fan can extract or deliver)

What is the pressure difference created by the fan? (determines the pressure drop, caused by resistance to air movement in the ducting, that the fan can overcome)

How much energy is the fan using? (determines cost of operating a ventilation system)

Without fans mechanical ventilation and air conditioning systems cannot operate. Optimal operation depends on a knowledge of the factors that affect fan performance.

The performance of fans in use is predicted by a set of laws which govern them. These laws can be placed into two groups; those that predict the changes arising from varying the speed of the fan and those which predict what happens when varying the size (diameter) of the fan (see also p64).

SPEED CHANGES (size kept constant)

Law 1. *Increasing the fan speed increases the volume of air drawn into the fan.* This is a direct relationship, so for example, doubling the speed of the fan doubles the air volume drawn into the fan.

Law 2. *The pressure difference generated across the fan inlet and outlet increases with increasing fan speed.* This is a square relationship so doubling the fan speed will quadrouple the pressure difference created by the fan.

Law 3. *The energy used by the fan increases as the fan speed increases.* This is a cubic relationship so doubling the fan speed will increase the electrical use by a factor of eight.

SIZE CHANGES (speed kept constant)

Law 1. *As the fan size increases the volume of air drawn into the fan increases.* This is a cubic relationship so doubling the fan diameter increases the input by a factor of eight.

Law 2. *The pressure difference created across the fan increases as the fan size increases.* This is a square relationship so doubling the size of the fan increases the pressure difference by a factor of four.

Law 3. *As the fan size is increased the energy required to operate it increases.* This is a cubic relationship so doubling the fan size will increase the energy consumption by a factor of eight.

Manufacturers produce fan characteristic curves (figure IP11) which graphically illustrate changes which occur as the volume flow rate changes. This characteristic is for an individual fan.

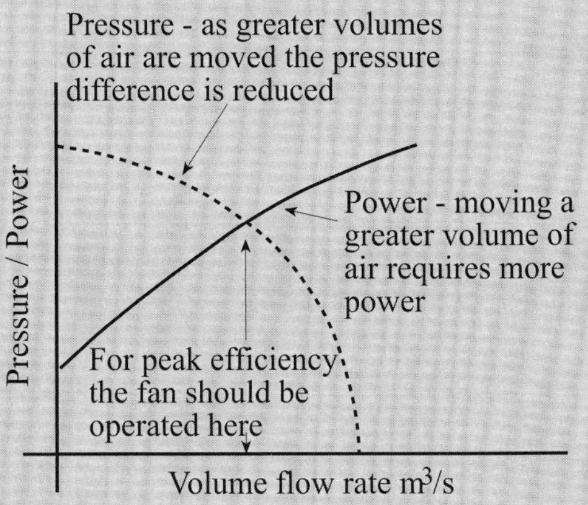

Figure IP11. Fan characteristic (radial bladed centrifugal fan)

sents a source of heat loss from the building. Mechanical ventilation has an advantage over natural ventilation in that systems can be put in place to recover most of the heat normally lost along with the extract air. The systems are known as air to air heat recovery units. Each method of heat recovery involves transferring energy from the exhaust airstream to the supply airstream. There are a number of ways of achieving this as described in the following sections.

Air to air plate heat exchangers are used for both domestic and commercial heat recovery. The remaining systems, because of their cost and complexity, are restricted to commercial use.

Plate heat exchangers are used where the exhaust and supply airstreams are arranged to flow alongside each other (figure 4.5). They are composed of a cubical sandwich of thin metal or plastic plates. These plates allow the exhaust and supply airstreams to pass each other but remain separated. Heat passes from the hotter to the cooler airstream by conduction through the thin plates (figure 4.6).

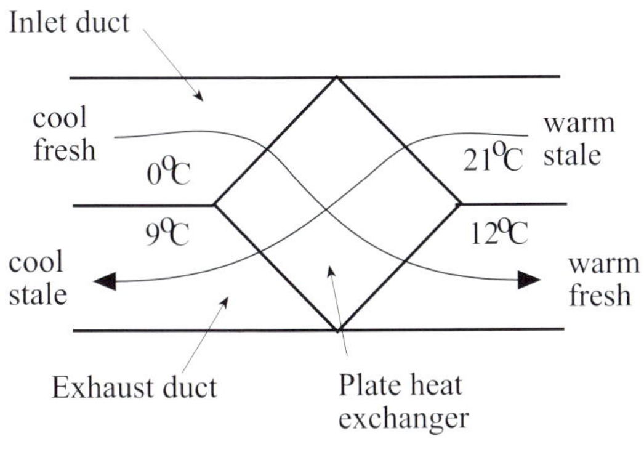

Figure 4.5 Location of plate heat exchanger in the ducting

Plate heat exchangers have a number of advantageous features;

- They have no moving parts which would require maintenance.

- They keep the airstreams separate so no cross contamination can occur.

- No energy is required for their operation although fan power may need to be increased to overcome air friction through the unit.

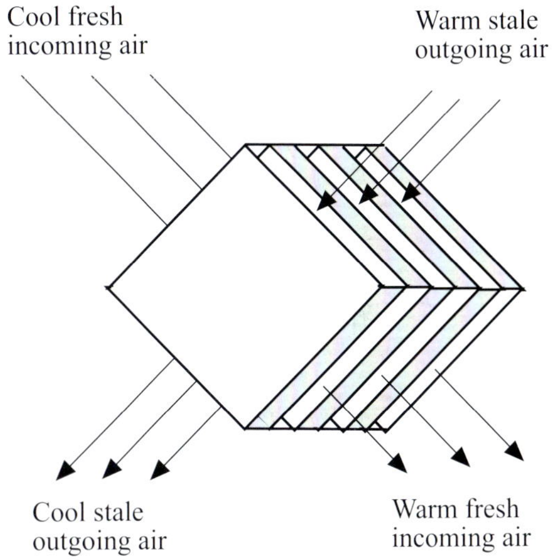

Figure 4.6 Air to air plate heat exchanger

Thermal wheels are composed of a circular matrix of tubes through which air can flow (figure 4.7). The wheel is positioned across the inlet and exhaust ducting so that inlet air passes through the upper half of the wheel and exhaust air passes through the lower half. As the exhaust air passes through the wheel the matrix heats up. The wheel rotates, and slowly bring this heated section into the path of the incoming airstream. The incoming airstream becomes heated by contact with the warmed thermal wheel matrix.

The thermal wheel requires an electric motor to drive it and so the energy consumption of this needs to be considered in assessing the heat recovery efficiency. Thermal wheels should not be used in areas where cross contamination of airflows would be a problem such as hospital operating theatres. This is because it is not possible to fully seal between supply and exhaust airflows.

Run around coils are heat recovery devices which can be used when supply and exhaust airflows are not run close together. A finned coil is situated in the path of the exhaust air (figure 4.8). Air passing through the coil heats up a water and antifreeze mixture circulating

IP12 - ECONOMICS.OF.HEAT.RECOVERY

Heat recovery devices represent a capital cost which must be recovered in the value of energy savings made. Economic viability usually depends on recovering this capital cost within a certain period of time known as the payback period. Determination of payback periods involves a comparison between capital and running costs on the one hand and value of energy savings on the other.

It is not always economical to use heat recovery devices. Capital and running costs must be weighed up against the value of energy saved.

Simple payback period calculations can be made using the following formula. For retrofitting of heat recovery devices a simple payback period of three to four years is usually considered economically viable.

$$\text{Payback period(y)} = \frac{\text{Capital Costs (£)}}{\text{Energy savings (£/y) - running costs (£/y)}}$$

PAYBACK FACTORS

Initial costs involve the capital and installation costs of the heat recovery device. However, some of this may be offset by savings arising from reductions in the size of boiler plant, made possible by the availability of recovered heat.

Running costs involve a debit in terms of electricity used by fans, pumps and motors and also maintenance costs. The credit is in the value of recovered energy. Running costs and credits are strongly dependent on the hours run by the system and availability of energy for recovery. For example cost effectiveness will be greater in winter when differences between inside and outside temperatures are at their greatest. Similarly the cumulative value of energy savings will be greatest in buildings where the ventilation system runs for a large number of hours in the day.

System efficiency varies depending on the rate of flow of air through the device. Figure IP12 shows how the efficiency of a domestic air to air plate heat recovery unit depends on the airflow rate. As the airflow rate increases the heat recovery efficiency decreases. This is because the air does not dwell in the device sufficiently long for heat transfer to take place. It can be seen that the peak efficiency is 75% indicating that three quarters of the heat contained in the extract air can be recovered to the supply air .

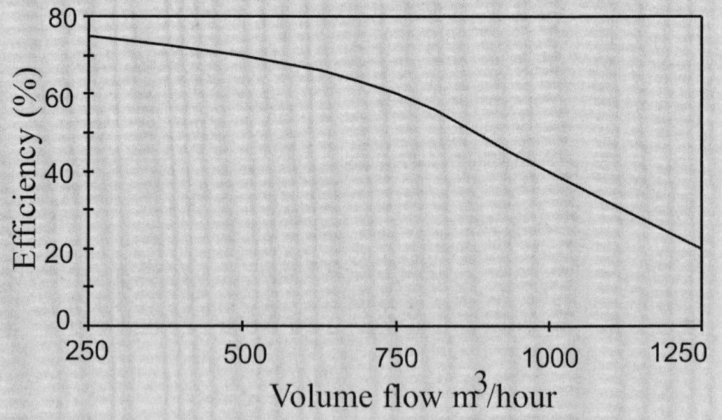

Figure IP12 graph of efficiency against airflow rate

ENVIRONMENTAL BENEFITS

Falling energy costs following privatisation of the fuel utilities means that it is increasingly difficult to justify energy saving devices in terms of financial payback. This is because the value of recovered energy has fallen. There are, however, considerations to be made beyond financial paybacks. Energy usage represents a cost to the environment. Heat recovery will result in a reduced need for heat generation and hence a reduction in the emission of pollutant gases such as carbon dioxide. Unfortunately no formal carbon payback methods exist.

in the pipe work. A pump circulates this heated solution to a similar coil installed in the supply duct. The supply air will become heated by passing through this coil. Topping up and maintenance of the system will need to be considered in economic analyses.

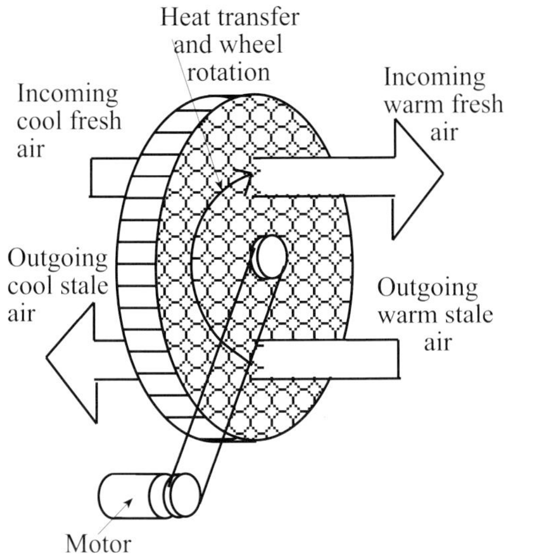

Figure 4.7 Thermal wheel

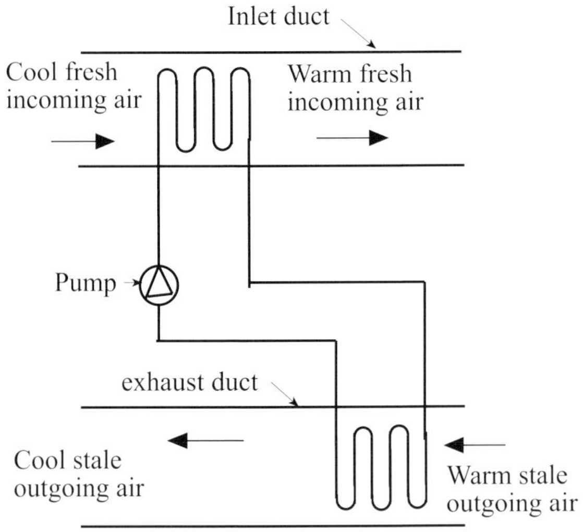

Figure 4.8 Run around coil

Heat pumps are similar to run around coils in that they can exchange heat between ducts which are separated by distance. However the pipework is filled with a refrigerant which evaporates in the coil situated in the exhaust duct. As the refrigerant evaporates it absorbs heat. The refrigerant vapour then flows to the coil in the supply duct where it condenses and in doing so releases the heat it has absorbed. The pipe work is fitted with a compressor and pressure reducing valve which enable the evaporation and condensation of the refrigerant to take place. A more full description of heat pumps is given in section 5.1.1.

Passive cooling

Mechanical cooling systems use energy to modify the ambient air temperature. It is sound practice to minimise heat gains to buildings and to use the fabric and natural sources before resorting to mechanical means of cooling.

Avoiding unwanted heat gains

The main sources of heat are the sun, people, office equipment and lighting. These sources provide typical heat input rates to offices of 60, 10, 25 and 25 W/m^2 respectively.

The sun

Too much glazing to the south, and especially the west facades should be avoided whilst maintaining sufficient for good daylighting. Special glazing can be used to prevent the heat from the sun entering at times that could cause overheating. Shading devices to the south can prevent sunlight entering during the summer, but still allow the heat from the sun to enter during the winter months when it is lower in the sky and the heat is desirable.

Equipment

The use of energy-efficient equipment should be encouraged.

Lighting

Lamps convert electricity to light but they also produce heat. The more efficient the lamp the more light and less heat is produced. Lamps should therefore have as high an efficiency as possible. In addition, the lamps should only operate when required.

Passive cooling

The simplest form of passive cooling is to ventilate the building with fresh outside air. So long as the outside air temperature is lower than the internal temperature the introduction of outside air will have a cooling effect.

The most commonly used passive cooling system is to use exposed mass within the space. An example of this is to leave the concrete ceiling exposed. Heat contained in the room air is absorbed by the mass reducing the air temperature. The heat gained by the fabric must be purged during the night so that it can act as a passive cooling mechanism again the next day. This is achieved by ventilating at night-time with cool air.

Further information

For further information on building-related energy efficiency, all it takes is one free phone call. Take action.

Call Action Energy today.
0800 58 57 94
www.actionenergy.org.uk

CARBON TRUST
Making business sense of climate change

AIR-CONDITIONING

5.0 Introduction

Air conditioning is the process by which the air in a space is modified to make it comfortable for the occupants. The primary function of air conditioning is cooling although all systems filter the air and some also provide heating and adjustments to the humidity levels.

Cooling is needed when the room air temperature rises above a comfort threshold of 27°C. Temperatures rise above this level due to a combination of high outside temperatures and internal heat gains. For example, in summer the outside air temperature may be 22°C or above. When this warm air enters the building its temperature will be further increased by heat gains from people, artificial lighting , appliances and the sun. Increases of 6°C due to these casual gains are not uncommon pushing the incoming air temperature above the comfort threshold. Even in winter when outside air temperatures are low, office buildings may experience sufficiently high casual heat gains that cooling is required.

Many of the situations previously described in section 4.2 as requiring mechanical ventilation also need a degree of air conditioning. To summarise, those situations requiring air conditioning are;

- Rooms subject to high solar gains, such as south facing rooms especially those with large areas of glazing

- Rooms with high equipment densities such as computer rooms and offices which make extensive use of IT

- Rooms in which environment (temperature, dust or humidity) sensitive work is being carried out such as operating theatres and microprocessor manufacturing units.

Air conditioning systems can be categorised into three main types;

Local comfort cooling systems - These systems cool the air in a room to bring its temperature down to acceptable levels. The cooling equipment is located in the room itself. The main forms of local comfort cooling system are;

- Window sill air conditioners
- Split systems
- Multi split systems
- Variable refrigerant flow split systems

Centralised air systems - All of the heating or cooling is carried out in a central air handling unit. Room by room control of temperatures is achieved using the following systems;

- Constant volume systems
- Variable air volume (VAV) systems
- Dual duct systems

Centralised air systems do not just provide heating or cooling but can filter, humidify or dehumidify the air as required. The central plant is usually in a ground floor plant room or may be a packaged unit situated on the rooftop.

Partially centralised air/water systems - A central air handling unit is used first to filter and then heat or cool an airstream. Final adjustment of temperatures is carried out using room based equipment. System types are;

- Terminal re heat or fan coil systems
- Induction systems
- Chilled ceilings and displacement ventilation

See IP16, page 80 for a basic system selection tree. All of the above systems and their components will be discussed more fully in later sections. The next

IP13 - LATENT.HEAT.RECOVERY.USING.HEAT.PUMPS

Water based leisure complexes by their very nature have large areas of water exposed to the air. The pool water readily evaporates increasing the relative humidity of the atmosphere. This in turn leads to thermal discomfort (see IP5) and damage to the pool structure due to corrosion and condensation of liquid water on cold surfaces. Evaporation can be slightly suppressed by keeping pool hall temperatures two degrees higher than pool water temperatures i.e. 29 and 27°C respectively and by keeping pool side air relative humidity levels between 60 and 70% RH. Increased activity in the pool or the presence of features such as flumes will increase the evaporation rate.

Evaporation of water from swimming pools creates a high relative humidity atmosphere which causes discomfort and damages the structure if left unchecked.

In the past problems were avoided by simply extracting the humid air out of the pool hall. This however, causes wastage of both energy and water. Energy wastage takes two forms firstly the sensible energy used to heat up the pool air and secondly the latent energy contained in the water vapour (see IP6). This latent energy is lost from the pool water itself during evaporation (approximately 0.7 kW per litre) and causes a reduction in temperature. If this energy cannot be recovered from moisture in the air, additional heat input from the boiler system will be required. It should therefore be a target of any design that latent energy should be recovered. Latent heat is released, and so can be recovered using a heat pump dehumidification system which condenses the water vapour back to a liquid.

An additional problem with pool hall air is that it contains chloramines. This is pollution, giving pools their characteristic odour, arising from the water treatment system and the bathers themselves. A fresh air supply is necessary to reduce the concentration of chloramines by dilution.

Water conservation is increasing in importance. Supplies are becoming more scarce resulting in shortages and rising prices. This has been brought about by increases in consumption and hotter summers created by global warming. It therefore follows that we can no longer be profligate in the use of water and that conservation is necessary. A heat pump dehumidification system can aid this by recovering up to 120 litres of water per hour from the extract air.

HEAT PUMP DEHUMIDIFICATION

Figure IP13 shows a schematic of a heat pump dehumidification system (also refer to section 7.5). The aim of this system is to extract warm humid air from the pool, remove sensible and latent heat from it and use this heat to re heat the pool water and the mix of fresh and dehumidified air supplied to the pool hall. Top up air heating is provided by a lphw heating coil supplied by gas boilers or a CHP unit (section 1.2.1).

Finally substantial savings in energy and reduced evaporation are achieved by using pool covers when the pool is not being used.

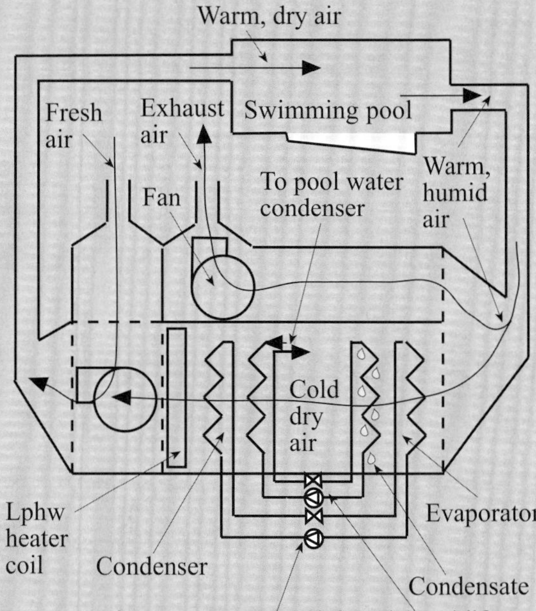

Two stage heat pump, pre cool then dehumidify increses energy recovery efficiency by 20%

Figure IP13 Heat pump heat recovery ahu

section will consider the cooling equipment on which all air conditioning systems are based.

5.1 Cooling

To warm air or water energy in the form of heat must be added to it. The converse is also true, to reduce the temperature of air or water energy must be removed from it. The system which is used by the majority of air conditioning systems is based on the vapour compression cycle. A less common system is absorption chilling (section 5.2)

It should be noted that the term "cooling" usually relates to the direct production of cold air whereas the term "chilling" relates to the production of cold water. This cold water is then used in the cooling coil of an air handling or fan coil unit to cool the airflow.

Vapour compression cycle. Most people have daily contact with cooling caused by the vapour compression cycle in the form of the domestic refrigerator. The refrigerator is a useful example to keep in mind whilst considering how the system works. Cooling systems used in buildings use the same principle but on a different scale. Figure 5.1 shows the components of a vapour compression chiller.

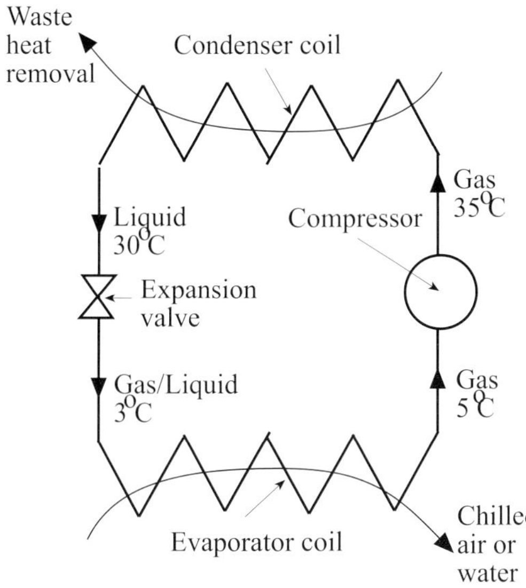

Figure 5.1 Vapour compression chiller

The main components are an evaporator coil, a compressor, a condenser coil, and an expansion device. These components are connected together using copper pipe through which refrigerant circulates in a closed loop. Cooling is achieved in the following way;

Liquid refrigerant is forced through the expansion valve. As the refrigerant leaves the expansion valve its pressure is reduced. This allows it to evaporate at a low temperature. For any liquid to evaporate it must absorb energy. The refrigerant evaporates by removing energy from the evaporator coil which in turn removes heat from the air which is flowing over it. Hence the air becomes cooled. The refrigerant, now in a vapour state, leaves the evaporator and passes through the compressor. The pressure is increased causing the refrigerant vapour to condense in the condenser coil. This occurs at a relatively high temperature. As the refrigerant condenses it releases the heat it absorbed during evaporation. This heats up the condenser coil. Air passing over the condenser coil takes away this waste heat.

In terms of a domestic refrigerator the evaporator would be situated in the ice compartment and the condenser is the grid of piping at the rear of the refrigerator which is warm to the touch. In building cooling systems significant amounts of waste heat are produced at the condenser and various techniques are used to safely remove it from the building. The method of heat rejection depends on the amount of waste heat produced and operational decisions such as the choice between using a dry system or a wet system. Air cooled condensers are discussed together with condensers utilising water in section 7.3.

The evaporator and condenser coils are simply arrays of copper pipe with aluminium fins mechanically bonded to their surface to increase the area for heat transfer. The following sections will consider refrigerants and compressors which are two of the more complex components of vapour compression chillers.

Refrigerants. Refrigerants are liquids that evaporate very easily at relatively low temperatures. Refrigerants are so volatile that if a liquid refrigerant was spilled in a room at normal temperatures it would very quickly evaporate away. Refrigerants must posses good thermodynamic properties but also have low toxicity and low flammability. Refrigerants are also the subject of

IP 14 - CAREFUL.USE.OF.REFRIGERANTS

It is estimated that 20% of refrigerant based systems will develop a leak resulting in a complete loss of refrigerant charge during their operational lifetime. Depending on the type of refrigerant the risks from this vary from fire and toxicity to global warming and ozone depletion. There are a number of methods of minimising or avoiding these problems. These include reducing the volume of refrigerant in the equipment, developing benign refrigerants, good practice including servicing, designing systems to avoid leaks and detection and rectification of leaks.

Some refrigerant gases are known to damage the ozone layer. However, they can only do this if they are allowed to escape from the system.

REDUCTIONS IN REFRIGERANT CHARGE

If a leak occurs the amount of damage caused depends on the volume of refrigerant that has escaped. It follows therefore that if the volume of refrigerant used to charge the system can be reduced the effects of a total leak will be minimised. This can be achieved in a number of ways. The first method is to reduce dependance on refrigeration equipment. This can be achieved using passive cooling techniques. This may eliminate the need for refrigerants completely or reduce the size of the system required. The second involves chosing equipment that has a high efficiency. This means more cooling can be carried out with less refrigerant. Hydronic systems can be used where the chilled water is created in a local plantroom. This is then used in the building rather than use refrigerant pipework through the building which would increase the the number of refrigerant components through which a leak could occur.

LEAK DETECTION

Refrigerant leak detection takes two forms, visual and gas analysis. Visual systems require a fluorescent dye to be added to the refrigerant. If the refrigerant begins to leak out of the system say through a loose joint then this will be revealed under an ultra violet light as a glowing patch of dye. In this way the exact location of the leak can be pin pointed. Unfortunately the leak can only be detected if the system is inspected regularly. The second type of leak detection involves drawing a sample of air surrounding the refrigeration equipment into a gas analyser. The analyser will detect and warn of the presence of refrigerant in the air sample indicating that a leak was occuring. This system can be set up to continuously monitor a plant room for the signs of a leak. Pin pointing the leak would require a further inspection of the system using either a hand held detector or searching for the presence of escaped dyes as described above.

RECYCLING OF REFRIGERANTS

The production of a number of refrigerants has been banned under the Montreal Protocol and EU regulations. However existing stocks can still be used to service older equipment. It follows therefore that the refrigerant contained in systems about to be replaced has considerable value to existing users. It is illegal under the environmental protection act to release substances into the environment which are known to cause damage. Because of this all refrigerant should be removed from the system and stored before repair or decommissioning.

If the recovered refrigerant is of good quality it can be re used without further treatment. If the refrigerant is contaminated with oils, acids, moisture or particles then the refrigerant must be cleaned by filtration and distillation before being re-used. Heavily contaminated refrigerants must be reclaimed this requires that they are taken off site and purified to their original state.

A good network of refrigerant reclaimers and recyclers is important to manage refrigerants and deter the growing trade of smuggling illegal refrigerants into the country. See also IP15 - Refrigerants and the Environment.

environmental concerns as described in IP14 and IP15.The evaporation and condensation of refrigerants in a chiller is controlled by lowering and increasing the pressure using the expansion valve and compressor respectively.

Compressors are electrically driven pumps of which there are three main types. These are; reciprocating, rotary and centrifugal compressors. Reciprocating compressors work by allowing refrigerant to flow into a chamber on the down stroke of a piston. The refrigerant is then forced out of this chamber towards the condenser as the piston moves upwards once more, Rotary compressors have two interlocked helical screws which when rotated move refrigerant which is trapped between the two screws along the line of the thread. Centrifugal compressors have a rotating impeller which forces the refrigerant outwards against the casing. This force is sufficiently strong to drive the refrigerant towards the condenser.

Compressors are further classified as either hermetic, semi hermetic or open depending on the seals between the motor and the compressor it drives. Hermetically sealed compressors have the motor and compressor together inside a shell whose seams are sealed by welded joints. Refrigerant is in contact with the motor and compressor. Semi hermetic compressors are similar but the joints are bolted rather than welded allowing servicing to take place. Open compressors have an external motor connected via a shaft to the pumping mechanism. A seal around the shaft stops refrigerant escaping.

The different forms of compressor are suitable for different cooling load ranges. Reciprocating up to 180kW, Rotary up to 2MW and centrifugal in the range 180kW to 3.5MW. Comfort cooling tends to be of a lower cooling capacity and so uses reciprocating compressors. Rotary and centrifugal compressors are used for large capacity centralised cooling systems.

5.1.1 Heat Pumps

Heat pumps are vapour compression systems, as described previously, but they are used for space heating rather than cooling.

It can be seen that what the vapour compression chiller is doing is extracting heat from a low temperature space and transferring it into an enviroment at a higher temperature. This is the basis of the heat pump (figure 5.2) which uses the vapour compression cycle to absorb heat from outside air and convert it to higher grade heat for indoor space heating.

The theoretical efficiency with which the heat pump carries out this function is very high at approximately 300%. This means for every 1kWh of electricity put into the compressor 3kWh of heat is obtained by the building. In practice however the operating efficiency tends to be lower. This is for two main reasons. The first is that the highest efficiencies are obtained when the inside and outside temperatures are similar. This is not the case in winter when heat pumps are required for space heating. The second cause of the fall off in efficiency occurs on cold winter days when the evaporator may become iced up due to low temperatures. This restricts heat transfer across the evaporator. This can be avoided by using an electrical heater on the outside coil to defrost it or to reverse the refrigerant flow direction. Both of which reduce the overall efficiency of the device.

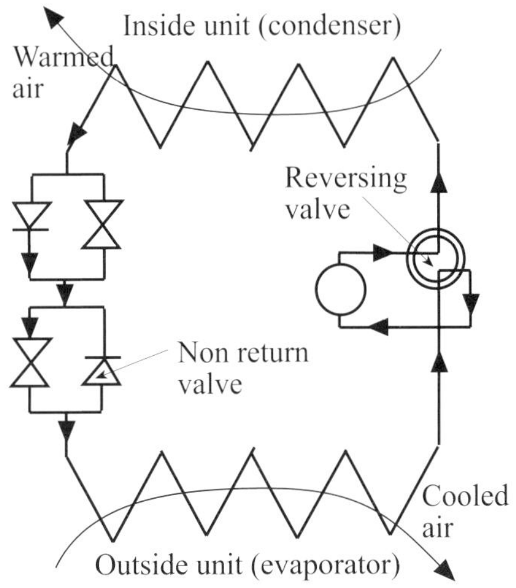

Figure 5.2 Reverse cycle heat pump in heating mode

Even with this reduction in efficiency the efficiency with which the unit uses electricity to provide heating is higher than simple resistive heating. The operating effi-

NOTES

ciency of the device can be increased if a body of water is used as the heat source rather than the outside air. This is because the water will have a more stable and higher temperature than the surrounding air. Examples are canals, lakes, ground water or warm effluent.

Reverse cycle heat pumps are very useful pieces of equipment which can either heat or cool a space. This feature is obtained by equipping the heat pump with a valve which can reverse the direction of refrigerant flow (figure 5.3). The direction of the refrigerant flow determines if the coil inside the building is a cooling evaporator or heating condenser. Two expansion valves fitted with non return valves are also required. Each expansion valve works in one direction only.

Reverse cycle heat pumps are particularly useful where spaces may have a requirement for both heating and cooling but at different times. One application is in shops where at the start of the day heating may be required. Later in the day as the shop fills with customers and heat is given out by display lighting, cooling may be needed to maintain comfort.

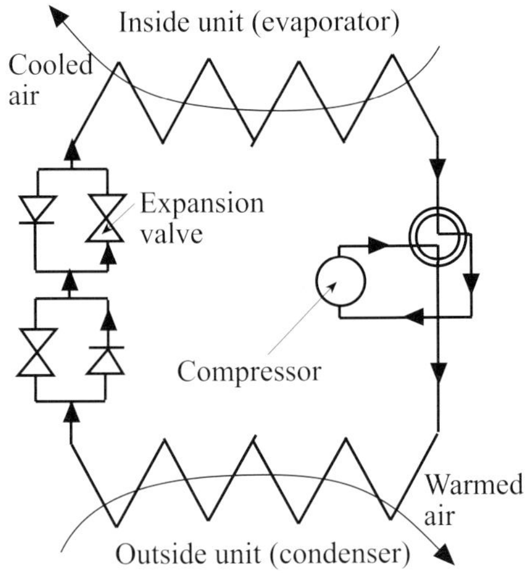

Figure 5.3 Reverse cycle heat pump in cooling mode

5.2 Absorption Chilling

There is growing interest in a method of cooling buildings which uses gas as a fuel instead of electricity. The technology is known as absorption cooling. The biggest differences between this and vapour compression cooling is that the compressor is replaced by a gas fired generator and the refrigerant is replaced by a refrigerant/absorber mixture. A diagram of an absorption chiller is shown in figure 5.4. The generator is filled with a mixture of refrigerant and absorber (solvent) which can be either water/lithium bromide (>35kW capacity) or ammonia/water. (>11kW). Note because water freezes the lithium bromide/water units can only cool down to 5°C, ammonia/water units on the other hand can cool down to -10°C. The way the system works can be illustrated using the ammonia/water pairing as an example. In this case water is the absorber and ammonia is the refrigerant. The water is called the absorber, giving the process its name, as it is so chemically attracted to ammonia vapour that it absorbs it out of the atmosphere.

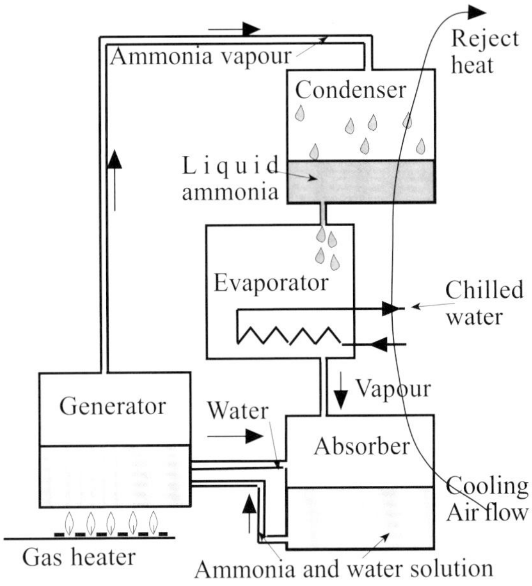

Figure 5.4 Absorption Chiller

A concentrated solution of ammonia in water is heated in the generator (figure 5.4) using a gas burner. The ammonia component vaporises first, as it has a lower

IP15 - REFRIGERANTS.AND.THE.ENVIRONMENT

During the last decade some refrigerants have been identified as ozone depleting gases and/or greenhouse gases. As a consequence the chemical companies producing refrigerants have been working to find alternative refrigerants which have a good blend of physical and thermodynamic characteristics but do not damage the environment if they escape. At the same time governments have brought in legislation which bans the production and use of the more damaging refrigerants. The most well known of these pieces of legislation is the *Montreal Protocol on substances which deplete the ozone layer*. This legislation has banned the production of the most ozone depleting refrigerants and has set a time limit on the manufacture of less damaging refrigerants. Many governments and the EU have brought in more strict legislation shortening timescales meaning that bans are now in place.

Some refrigerants have been identified as contributing to ozone depletion and global warming

Refrigerants are identified in the building services industry by a refrigerant number. For example, R11 and R12 are chlorofluorocarbons (CFC's) which are highly destructive to the ozone layer and their production is now banned. R22 is an hydrochlorofluorocarbon (HCFC) which is less damaging to the ozone layer than CFC's and so its production is allowed until 2005. Existing stockpiles of both refrigerants can still be used. R134a is a Hydrofluorocarbon (HFC) it contains no chlorine and so does not damage the ozone layer. However like other refrigerants it is a global warming gas.

There are three indicese that are used for comparing the environmental effects of refrigerants;

Ozone Destruction Potential (ODP) - A measure of how destructive the chemical is to the ozone layer in comparison to R11 which is said to have an ODP = 1

Atmospheric Lifetime - The length of time, measured in years, that the refrigerant remains in the atmosphere causing ozone destruction.

Global Warming Potential (GWP) - A measure of the contribution the chemical makes to global warming in comparison to CO_2 whose GWP = 1.0.

Table IP15 below compares these indecese for various refrigerants. It can be seen that R134a has a zero ODP but still has a global warming potential. Environmental groups are now campaigning against HFCs because of their GWP. However, the dominant factor in global warming is CO_2 emitted (from power stations) as a result of electrical consumption by the chiller rather than the global warming effect of escaped refrigerants. A method of quantifying the contributions from each is given by the Total Equivalent Warming Impact (TEWI). This is a lifecycle analysis which considers both the direct global warming impact of the escaped refrigerant and the efficiency of the refrigeration system as a whole.

Environmental concerns have also led to renewed interest in traditional refrigerants such as ammonia and propane. Both of these do not affect the ozone layer or add to global warming. There are concerns over toxicity and flammability of these refrigerants and so they should be used externally and according to appropriate guidelines. Absorption chillers which use a mix of ammonia/water (section 5.2) and waste heat which would otherwise be wasted have a low contribution on global warming and ozone depletion when compared to other systems.[1]

Refrigerant	Type	ODP	Lifetime	GWP
R11	CFC	1.0	60 years	1500
R22	HCFC	0.05	15 years	510
R134a	HFC	0.0	16 years	420
R290	Propane	0.0	<1year	3
R717	Ammonia	0.0	<1year	0
Lithium Bromide		0.0	<1year	0

Table IP15 Environmental Indecese

1. Phone the ETSU enquiries bureau 01235 436747 to obtain a free copy of Good Practice Guide 256: An introduction to Absorption Chilling

boiling point than water, and passes into the condenser. The water which is left behind passes back to the absorber. The ammonia vapour condenses back to liquid ammonia in the condenser giving out waste heat. This heat is removed from the system by air which is blown over the condenser by a fan (section 7.3). The ammonia now passes from the condenser into the evaporator via an expansion valve. In doing so its pressure drops and so it can evaporate once more. It does this by absorbing heat from the chilled water circuit. Chilling has therefore, been achieved. The ammonia vapour now passes into the absorber where it is absorbed by the water from the generator to create a concentrated ammonia solution. Heat is given out when the two chemicals combine. This waste heat is also removed by the condenser cooling air flow. The ammonia solution is pumped back to the generator where the cycle continues once more.

The above device is known as a single effect absorption chiller. Double effect units are also available which use water and lithium bromide. This solution is pre heated on its way back to the generator by passing it through a heat exchanger. This improves the efficiency of the unit. Double effect units require a higher temperature heat source (>140°C) derived from a direct gas fired burner or pressurised hot water.

Absorption chillers are less efficient than vapour compression chillers with a COP of approximately 0.7-1.2. It follows that more gas energy will be required than an equivalent electric chiller (COP = 3.0). However the cost and pollution differentials will be reduced because electricity costs and pollutes approximately four times more per unit of energy than gas because of wastage in the power stations. Contract gas prices are lower still in summer when gas is needed for cooling as less is needed for space heating.

As well as direct gas firing some absorption chillers can be operated using waste heat. One form of surplus heat is that generated by combined heat and power units. In winter their heat output is used for space heating. In summer this heat is surplus to requirements and so can be used to drive the absorption chiller. This is known as trigeneration or combined cooling and power. When heat, which would normally be wasted, is used absorption chillers emit much less CO_2 into the atmosphere than a vapour compression chiller for a given cooling effect (see IP15). Research is currently underway which is investigating the linking of absorption chillers with solar panels as a source of generator heat. It is an advantage that the appearance of large amounts of cost and pollution free solar energy coincides with the need for cooling.

OTHER BENEFITS

The only electrical elements in an absorption chiller are the pumps used to move the ammonia/water solution from the absorber back to the generator. These pumps consume much less power and produce less noise and vibration than a compressor. This latter point is useful if the chiller is to be sited near to a noise sensitive area.

The pump along with the air cooled condenser fan and the gas burner fan are the only moving parts. The rest of the device consists of sealed metal chambers. This configuration means that maintenance costs are low.

External, air cooled, modular packaged units mean that cooling capacity can be easily expanded as the building is developed or as heat loads increase. Flexibility is further enhanced as units are also available that provide heating in winter and switch to cooling in summer. Features and photographs of commercially available absorption chillers are shown on page 76.

SELECTION CRITERIA

From the above it can be seen that absorption chilling is particularly appropriate where;

- You have excess heat production from your CHP plant in summer or a production process which can be used to drive the absorption chiller
- The electrical supply to the site is not robust enough to supply the necessary electricity required for vapour compression chilling and an expensive upgrade would be necessary.
- You wish to optimise the use of clean gas as a fuel throughout the year, not just in winter
- You have a source of low cost or free heat energy available such as solar energy or heat released from the combustion of landfill gas.

IP16-VENTILATION.AND.AIR-CONDITIONING.SELECTOR

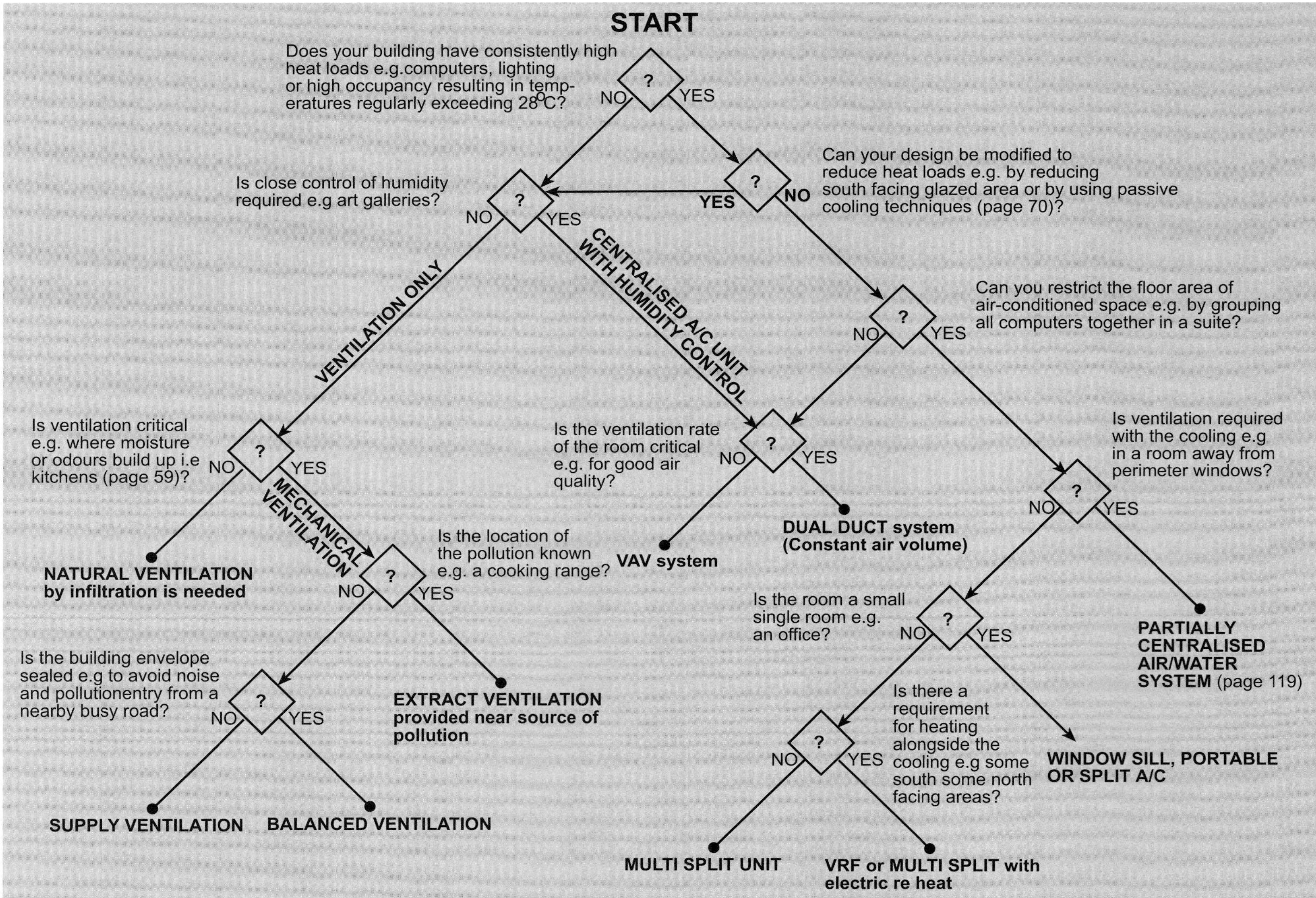

START
Does your building have consistently high heat loads e.g.computers, lighting or high occupancy resulting in temp-eratures regularly exceeding 28°C?
NO
YES
Can your design be modified to reduce heat loads e.g. by reducing south facing glazed area or by using passive cooling techniques (page 70)?
YES
NO
Is close control of humidity required e.g art galleries?
NO
YES
VENTILATION ONLY
CENTRALISED A/C UNIT WITH HUMIDITY CONTROL
Can you restrict the floor area of air-conditioned space e.g. by grouping all computers together in a suite?
NO
YES
Is ventilation critical e.g. where moisture or odours build up i.e kitchens (page 59)?
NO
YES
MECHANICAL VENTILATION
NATURAL VENTILATION by infiltration is needed
Is the location of the pollution known e.g. a cooking range?
NO
YES
EXTRACT VENTILATION provided near source of pollution
Is the building envelope sealed e.g to avoid noise and pollutionentry from a nearby busy road?
NO
YES
SUPPLY VENTILATION
BALANCED VENTILATION
Is the ventilation rate of the room critical e.g. for good air quality?
NO
YES
VAV system
DUAL DUCT system (Constant air volume)
Is ventilation required with the cooling e.g in a room away from perimeter windows?
NO
YES
PARTIALLY CENTRALISED AIR/WATER SYSTEM (page 119)
Is the room a small single room e.g. an office?
NO
YES
WINDOW SILL, PORTABLE OR SPLIT A/C
Is there a requirement for heating alongside the cooling e.g some south some north facing areas?
NO
YES
MULTI SPLIT UNIT
VRF or MULTI SPLIT with electric re heat

This ventilation and air-conditioning selector chart is for guidance only, It is intended to illustrate some of the issues involved in the selection process. There are other issues involved which are not considered here and which may take priority, these include; Capital and maintenance costs, Cooling capacity and Energy consumption. To be able to make comparisons, these later factors are stated in pounds or watts per square metre of treated floor space. It is likely that in the future the decision to air-condition a building will be subject to building regulation approval. For further details on selection criteria consult the CIBSE Guides or phone BRECSU on 0800 585794 and ask for a free copy of *Good Practice Guide 71: Selecting Air Conditioning Systems.*

6.0 Local Comfort Cooling Systems

Comfort cooling systems operate by circulating room air over the evaporator coil of a vapour compression chiller so that it becomes cooled. The system also includes a method of rejecting the waste heat from the cooling process outside of the building.

The vapour compression cycle is used in a number of commercial room cooling products. The main variants are; Window sill, split, multi split and variable refrigerant flow air conditioners.

Window sill air conditioners are the most basic form of cooling system. They are typically used as a retrofit solution to an overheating problem which may have arisen due to the introduction of computers into an office space. The refrigerating equipment is contained within a cabinet which sits on the window sill (figure 6.0).

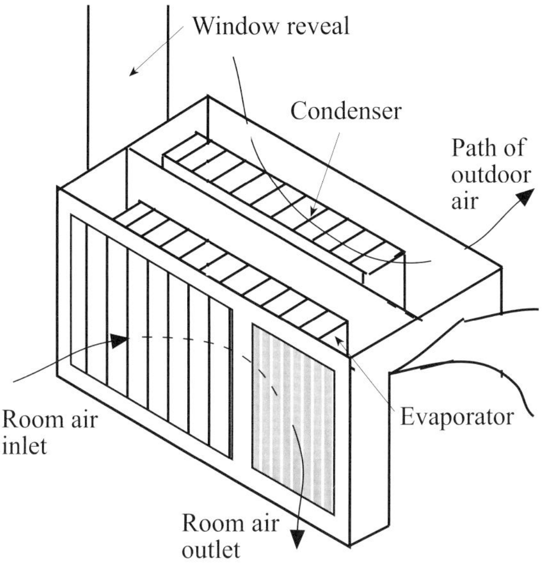

Figure 6.0 Window sill air conditioner

The window must be modified to seal the remaining gap above and to the sides of the unit. The room side of the air conditioner is sealed from the outdoor side. Air is drawn by a fan from the room, through a filter, over the evaporator coil and then is returned, chilled, back to the room. At the same time outside air is circulated over the condenser coil to carry away the waste heat to the outside. All the controls and the compressor are fitted into the casing to create a self contained unit.

Portable air conditioners are based on the same principle except that the cabinet is designed to be moved into different rooms as required. A length of flexible ducting which runs from the cabinet to the outside through an available opening such as a window is used to discharge waste heat out of the building.

Split air conditioning systems. Split air conditioning systems are so described because the evaporator is housed in a room unit and the condenser is housed in a separate outdoor unit. Refrigerant flow and return pipes connect the two units together. The indoor unit can be wall or floor mounted or accommodated within a suspended ceiling. The finish of the indoor unit is of high quality to integrated with the appearance of the room decor or suspended ceiling panels.

A diagram of a ceiling unit is shown in figure 6.1. A fan is used to draw room air across the evaporator to provide the necessary cooling. Chilled air is then output via directional slots. These slots are adjusted to keep the cold airstream away from the room occupants so that cold draughts are avoided. The chilled air mixes with the room air outside the occupied zone. The mixed air eventually diffuses throughout the room to create the cooling effect.

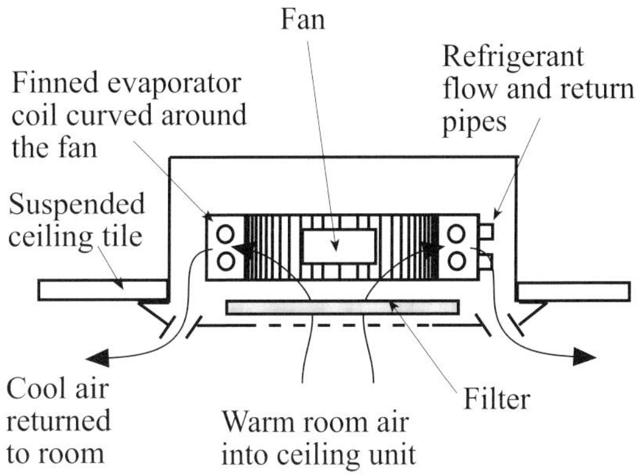

Figure 6.1 Indoor unit (cassette) of a split air conditioning system

NOTES

The outdoor unit (figure 6.2) contains the condenser which is air cooled. The condenser like the evaporator has its surface area increased using fins. A fan is used to draw outside air across the condenser to discharge the waste heat to atmosphere. The outdoor unit can be placed a considerable distance (up to 50m pipe length including 30m vertical rise) from the indoor unit. This allows flexibility of design and sympathetic positioning of the outdoor units on the external surfaces of the building.

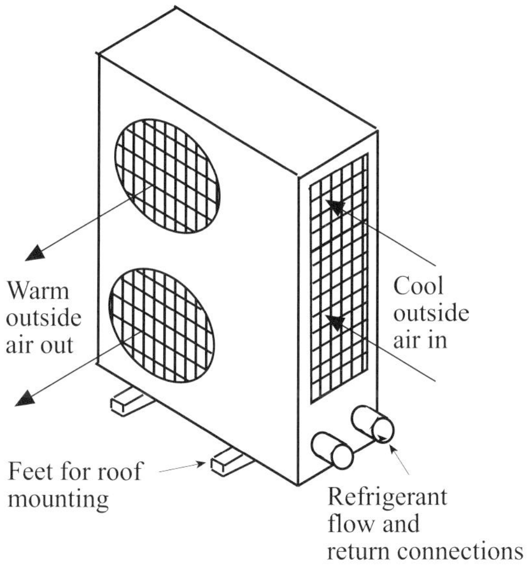

Figure 6.2 Outdoor unit of a split air conditioning system

Multi split air conditioning. Is based on the same principle as single split air conditioning except that up to four indoor units can be served by a single outdoor unit. Each indoor unit has its own set of refrigerant pipe work connecting it to the outdoor unit. All of the indoor units operate in the same mode i.e. all heating or all cooling, although individual control of the degree of heating or cooling from off to full output can be exercised over each unit. Some indoor units are fitted with electric heaters so that whilst the multi split group is operating in cooling mode odd single units can provide a degree of heating.

Variable refrigerant flow (VRF) air conditioning. In this system up to eight indoor units can be operated from a single outdoor unit. The main advantage of this system over multi splits is that each indoor unit can operate either in cooling or heating mode independently of the other units. This is achieved by having collection vessels for both vapour and liquid refrigerant (figure 6.3). A sophisticated control system redirects these two refrigerant phases to the indoor units as required. As a consequence VRF air conditioning systems incorporate heat recovery as part of their mode of operation. Waste heat say from rooms on the south side of the building can be re distributed by the refrigerant to indoor units on the north side of the building. The distance between indoor and outdoor units can be up to 100m including a vertical rise of 50m.

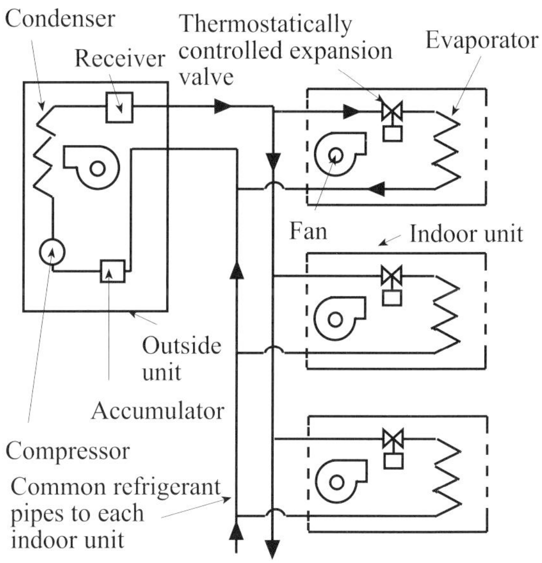

Figure 6.3 Variable refrigerant flow system

Filtered and tempered air can be supplied to each unit from a centralised air handling unit to provide ventilation as well as heating/cooling.

Water to air reverse cycle heat pumps are a heat pump system which gives the opportunity for efficient operation through heat recovery. The water to air reverse cycle heat pump system is also known as the versatemp system after the first commercial system produced by Clivet Ltd. The system (figure 6.4) is comprised of room based reverse cycle heat pumps. These heat pumps have an air coil which supplies heat or cooling to the room depending on the direction of refrigerant flow. The other coil is part of a refrigerant to

NOTES

water heat exchanger. The water flow and return pipes to this heat exchanger connect into common flow and return pipes which also serve other reverse cycle heat pumps throughout the building.

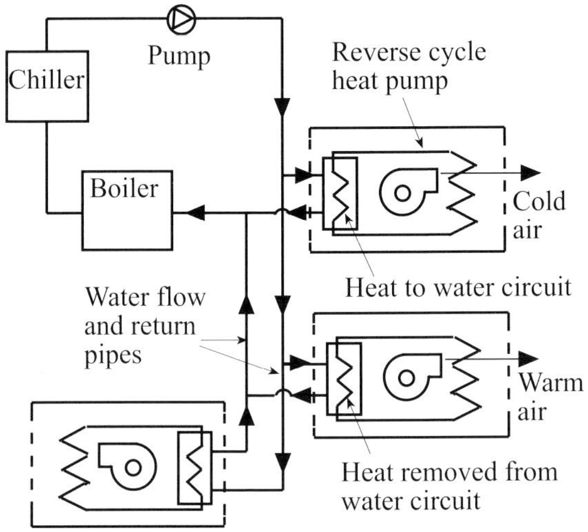

Figure 6.4 Water to air reverse cycle heat pump system

Heat pumps operating in cooling mode will extract heat from the room and deposit it into the water circuit. Other heat pumps which are in heating mode will take heat from the circuit. In this way the reject heat say from computer rooms can be recovered and deposited into rooms requiring heating. If it is required additional heating or cooling can be input to the water loop using boilers or chillers respectively.

Chilled water fan coil units. This system of comfort cooling uses a centralised chiller to produce cold water. This water is then distributed to room based fan coil units (figure 6.5). The fan coil units provide cooling in a similar manner to split system indoor units. The difference is that the heat transfer coil is filled with cold water instead of refrigerant. The units can be floor or wall mounted or recessed into a suspended ceiling.

There are a number of advantages to this system;

- Fewer constraints on the number of room based units

- The distribution system uses chilled water instead of refrigerant. As a result part of the system can be installed by tradesmen used to water systems as opposed to specialist refrigeration engineers.

- Chilling and heat rejection occurs in the centralised chiller. This may be in a plant room or on the roof top. The charge of refrigerant is therefore reduced and since the lengths of refrigerant pipe work are shorter the risk of leakage is diminished.

- Use of hydronic circuits and low fan speeds results in quiet operation making the units useful for noise sensitive locations.

Supplying the chilled water fan coils with fresh air using ducting brings the system closer to a partially centralised air/water system as described in section 8.0

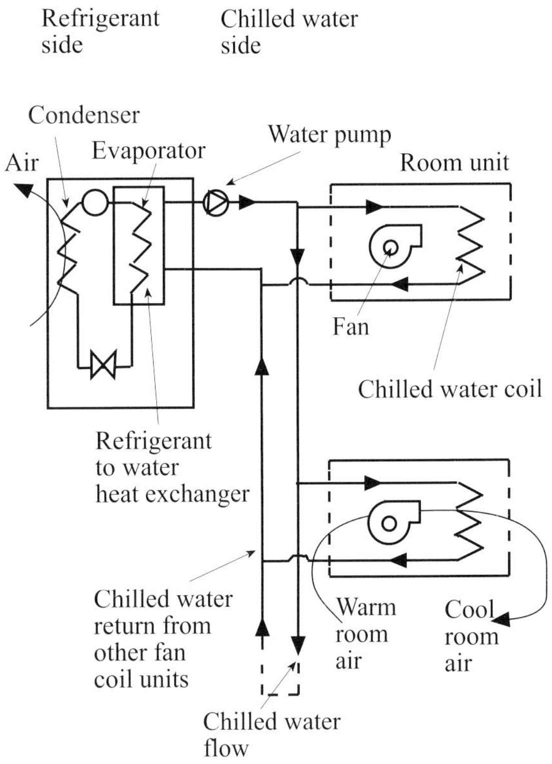

Figure 6.5 Chilled water fan coil units

IP17-CENTRALISED.A/C.SYSTEM-MAIN.COMPONENTS

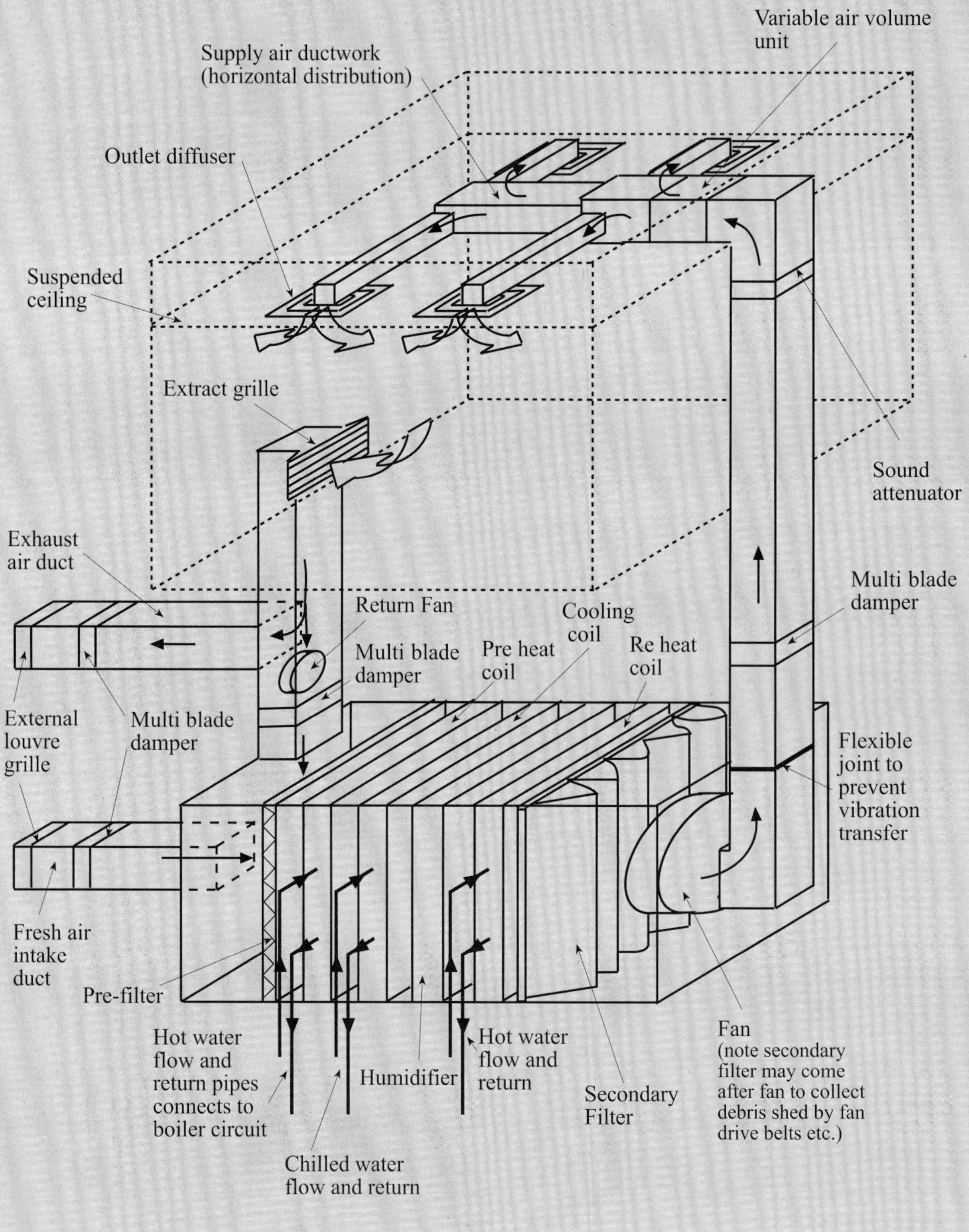

7.0 Centralised Air Conditioning Systems

Centralised air conditioning systems differ from comfort cooling systems described previously in that they are able to humidify or dehumidify the airstream in addition to providing cooling, heating and filtration. These changes are applied to the air using an air handling unit situated in the plant room or enclose on the roof. The conditioned air is then delivered to the rooms using ducting.

At the heart of a centralised air conditioning system is an air handling unit (AHU) (figure 7.1and IP17). This is a pressed steel cabinet containing the various components needed to condition the air which passes through it. Air is brought into the air handling unit via an inlet grille built into an external wall. This should be located to avoid sources of dust and pollution such as near by roads. If cooling the building is a priority then a north facing inlet grille will provide cooler inlet air temperatures. Rooftop inlets are often used in cities to avoid ground level pollution. Air enters the AHU where it is suitably conditioned by passing through filtration, heat recovery, humidity control and chilling or heating stages. A centrifugal fan drives the air movement through the AHU.

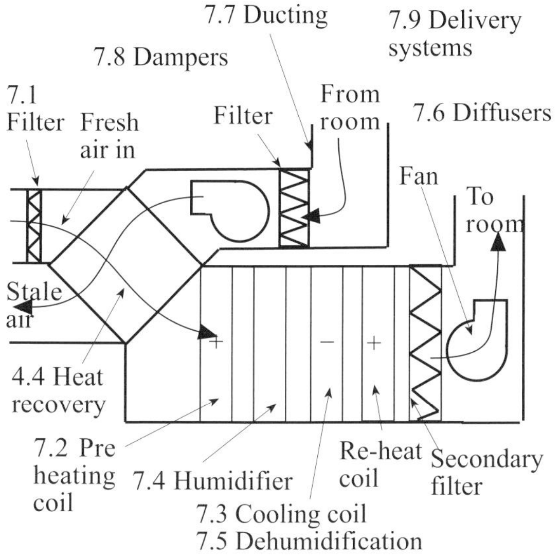

Figure 7.1 Air handling unit including heat recovery

Centralised air conditioning systems must have some way of responding to changes in demand for heating or cooling from the spaces. This is achieved in the way that conditioned air is delivered to the rooms. The methods used are; constant volume systems, variable air volume (vav) systems and dual duct systems. Each of these will be discussed in section 7.9 following a description of the main components in a centralised air conditioning system.

Here we will be discussing components in relation to a centralised air conditioning system but it should be remembered that many of the components are also used in other systems. For example filters and heating/cooling coils are used in fan coil units (section 8.0) and ducting is used whenever air movement needs guidance such as in extract ventilation systems (section 4.2).

7.1 Filtration

Air carries with it a large quantity of suspended particles including dust, fibres, bacteria, mould and fungal spores, viruses and smoke. It also carries gaseous pollutants such as benzene, NO_2, SO_2 and other odours emitted from vehicle exhausts. This is especially so where buildings are sited in urban areas where there is typically twenty times as much dust per cubic metre of air than is found in rural areas. The particles in the airstream vary in size from the visible such as hair and ash to the microscopic such as bacteria and viruses (IP18). It is essential that air passing through an air handling unit is filtered to remove these impurities. Inadequate removal of these particles leads to problems of poor air quality and ill health. Dust particles accumulate on heating/cooling coils reducing their effectiveness. Ductwork surfaces become coated in dust providing a breeding ground for bacteria. Finally, the presence of dust in the airstream leaving an air outlet causes unsightly dirty streaks on adjacent surfaces. In 1996 health and safety regulations concerning ventilation system maintenance and cleaning came in to force requiring that ductwork be cleaned on a regular basis

There are two methods by which air can be filtered these are mechanical filtration and electrostatic filtration. Gases and vapours are removed from the

NOTES

airstream using the process of adsorption.

MECHANICAL FILTERS

Mechanical filtration involves passing the airstream through a porous material known as the filter media. The materials are usually fabrics, glass fibre, non-woven synthetic materials or paper. Each is held across the airstream by a supportive framework. The capture process involves three mechanisms; Direct interception, inertial impaction and diffusion. These involve either directly stopping large particles as a result of a single collision or by gradually slowing down smaller particles by multiple collisions with successsive fibres. Eventually the small particle loses energy and comes to rest. A fourth capture mechanism for some filters is to give the fibres an electrical charge during manufacture. This will attract dust out of the airstream but the effectiveness will reduce with time as the charge is lost.

Since filters are required to be changed regularly it makes sense to construct them in a robust and effective way but cheap enough to be disposed of at the end of their life. It also makes sense to utilise the grades of filters in such a way as to extend the life of the most expensive filters for as long as possible. The effectiveness of lower grade (G2-G4) (see IP18) pad and panel filters can be enhanced by using a filter media of graduated density. This means that the back of the filter will have smaller pores than the front, and therefore will be able to capture a broader range of particulate sizes.

Pad Filters are used as the first bank of filters to prevent large particles from entering the system. A single flat sheet of material, they are held in a card or re-usable steel or lightweight aluminium frame. They protect the higher grade, and hence more expensive filters, next in line. They are normally referred to as Primary Grade Filters ranging from G2 to G4 in classification.

Panel filters. As shown in figure 7.2, the filter media is folded into pleats. This extends the media surface area when compared to the flat pad of material in a pad filter. The filter media is backed by an open strengthening grid and is then sandwiched into a card, wire-mesh or plastic frame. This panel can be easily slid into position in a metal holding frame in the air handling unit (AHU). They are particularly suited to AHU's which have insufficient depth to accommodate a bag filter (see later). When the filter requires changing it is removed, disposed of according to regulations, and a replacement inserted in the same manner. Some filters of this type are available as primary filters (usually G4) but more commonly are used for grades F5-F8 and as a secondary filter.

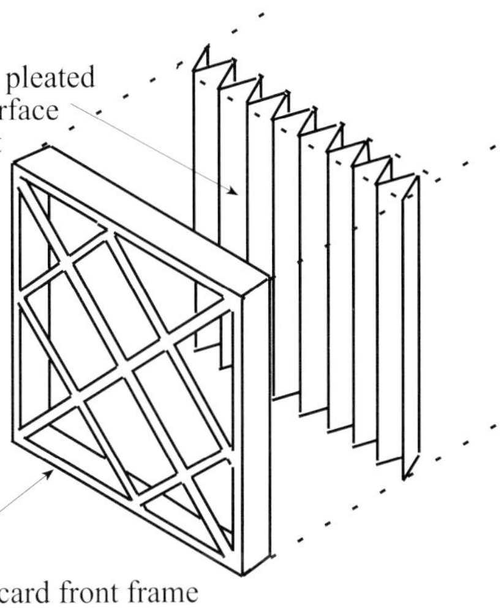

Filter media pleated to extend surface area for dust collection
Disposable card front frame a similar back frame is glued to this to hold the filter media in place

Figure 7.2 Panel filter

Anti-microbial filters. This is a panel filter surface sprayed or impregnated with biocides. The biocides used fall into one of two categories: inhibitor or eradicator. Inhibitors simply prevent the micro-organism from reproducing an eradicator kills it completely. The range of micro-organisms against which the biocide will be effective includes various types of bacteria, algae and yeasts. Anti-microbial filters are most commonly used in hygiene sensitive areas such as hospitals or food processing outlets. They are however being increasingly used in office environments to improve general indoor air quality and health. All grades of filter are available from G2 to F9.

Bag Filters have a filter medium which is formed into a bag and held in place by a metal or plastic frame (figure 7.3). The seams are well sealed and it is mounted so that the open end of the bag faces the oncoming air-

IP18 - AIR.FILTER.CHARACTERISTICS

The standard method of testing to evaluate performance of air filters in general ventilation and air conditioning is BS6540 (EN779:1993, European Standard). The test consists of two parts:

A: The synthetic dust weight arrestance test - providing an arrestance value when the filter is fed with a blended synthetic dust.

B: The atmospheric dust spot efficiency test - giving an efficiency value produced using an atmospheric staining technique.

The following values are obtained

Initial efficiency -
The efficiency of the filter against carbonaceous staining contamination in its clean state.

Initial arrestance - the effectiveness of the filter against large particulate matter in its clean state.

Both efficiency and arrestance have average values which represent performance at the average condition of the filter through its life.

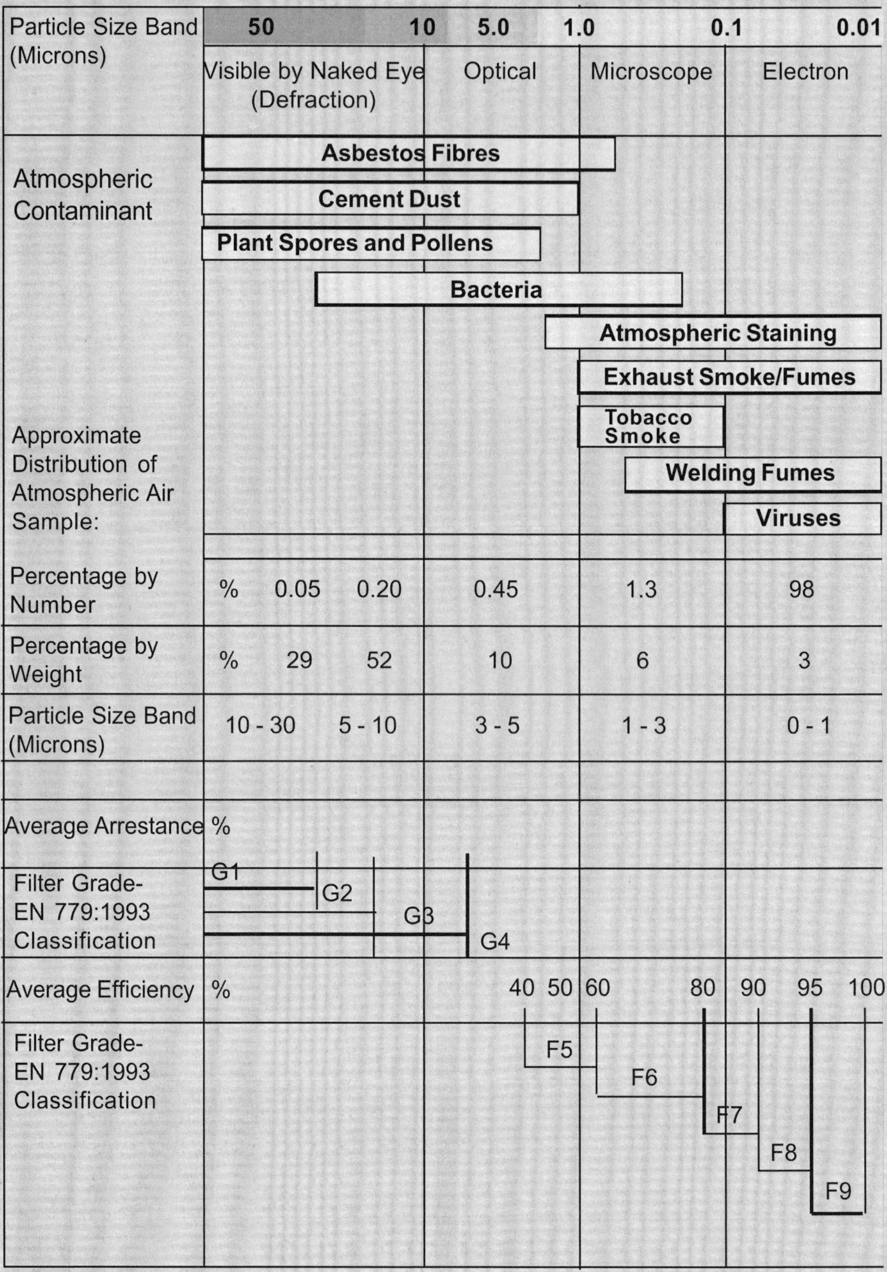

Particle Size Band (Microns)
50
10
5.0
1.0
0.1
0.01
Visible by Naked Eye (Defraction)
Optical
Microscope
Electron
Atmospheric Contaminant
Asbestos Fibres
Cement Dust
Plant Spores and Pollens
Bacteria
Atmospheric Staining
Exhaust Smoke/Fumes
Tobacco Smoke
Welding Fumes
Viruses
Approximate Distribution of Atmospheric Air Sample:
Percentage by Number
% 0.05 0.20 0.45 1.3 98
Percentage by Weight
% 29 52 10 6 3
Particle Size Band (Microns)
10 - 30 5 - 10 3 - 5 1 - 3 0 - 1
Average Arrestance %
Filter Grade- EN 779:1993 Classification
G1
G2
G3
G4
Average Efficiency %
40 50 60 80 90 95 100
Filter Grade- EN 779:1993 Classification
F5
F6
F7
F8
F9

Figure IP18 Dust particles, their sizes and appropriate filter grades

This chart has been reproduced with permission from publication NFC4 produced by the Nationwide Filter Company Ltd. 5 Rufus Business Centre, Ravensbury Terrace, London, SW18 4RL. Tel +44(0) 20 8944 8877.

flow. Providing that the bag has sufficient depth this form greatly extends the surface area over which filtration can take place and as a result bag filters have a long life, a high dust carrying capacity and offer a low resistance to airflow. To minimise the risk of sagging or collapse, some bag filters are manufactured with spacers to help the individual pockets remain open even at reduced airflows. Bag filters are usually available in grades from G4 to F9.

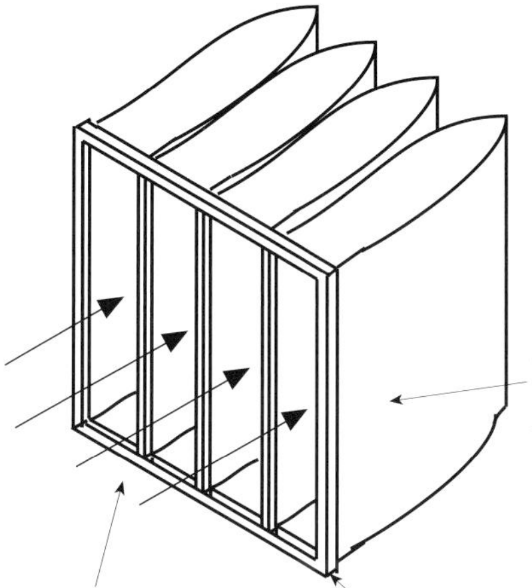

Figure 7.3 Bag filter

High Efficiency Particulate Air (HEPA) Filters are panel filters with extremely fine filter media with collection efficiencies ranging from 99.95 to 99.999%. To extend the life of a HEPA filter it must be used in conjunction with at least one pre-filter. HEPA filters may be included in the main area of the ahu or only at the air inlet grilles serving those rooms which require the cleanest air. HEPA filters are used where a very clean environment is required such as in micro electronics and pharmaceutical manufacturing and storage areas for documants and artefacts. An even higher grade of filter, the Ultra Low Penetrating Air (ULPA) filter is used in environments where ultra clean air is required such as the nuclear or space industries.

ELECTROSTATIC FILTERS

Electrostatic filters remove dust from the air by electrostatic attraction. Dust laden air entering the unit passes over an ioniser (figure 7.4). This induces a positive electrical charge on the dust particles. The airstream then passes between positively and negatively charged plates. The positive plates repel the charged dust particles towards the negative plates which are at the same time attracting the dust particles. The dust collects on the negatively charged plates.

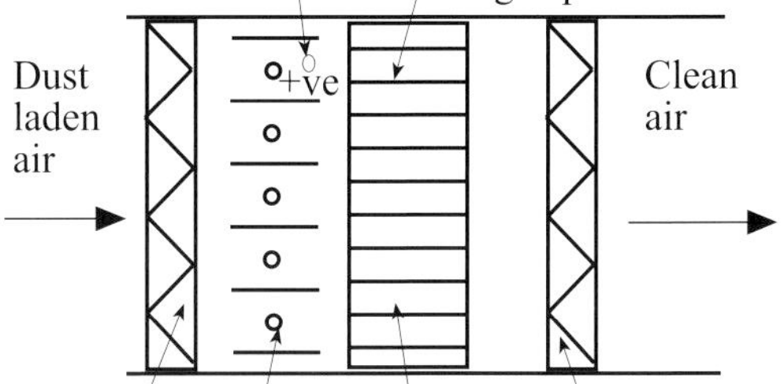

Figure 7.4 Electrostatic filter

Electrostatic filters have a mechanical pre-filter to remove the larger particles and a post filter to collect any large clumps of aggregated dust which may become dislodged from the unit. Some units have automatic cleaning systems which periodically wash down the collector plates which become coated in accumulated dust. In other systems the collector array is removed via a side hatch for cleaning before re assembly.

Electrostatic filters once seen as a low maintenance low pressure drop option have recently fallen out of favour. This is due to the increased cost of mechanical parts and the high cost of replacement plates that become less effective after 'pitting' and accumulation of inground atmospheric particulate staining.

ACTIVATED CARBON FILTERS

Activated carbon filters are used to remove gaseous pollutants and odours from the airstream which cannot be removed by mechanical or electrostatic filters. The carbonaceous material is first processed to produce a char and then heated to 800-1000°C to give it its micro-pore structure which enables the adsorption

IP 19 - MANAGEMENT.OF.FILTERS

There are two criteria used to compare the performance of filters. The first is the pressure drop which occurs from one side of the filter to the other. This pressure drop arises due to the resistance that the air encounters passing through the small pores in the media. The smaller the pores the greater the resistance. The second criterion is the filters ability to remove dust from the airstream, measured in terms of the filter efficiency. Various standard test methods exist (BS6540) (see IP 18) which involve measuring how much dust there is in the air upstream and downstream from the filter. The removal efficiency of the filter, expressed as a percentage can be calculated from these two values.

The efficiency will depend on the size of the dust particles and the pore size of the filter media. So for example a coarse filter with a relatively large pore size will have a high efficiency at collecting large particles but a low efficiency at collecting smaller particles. Materials with small pore sizes are good at removing both the large and small suspended matter. They do however, cause a greater pressure drop within the system. A large pressure drop will necessitate the use of a higher capacity fan which will increase the electrical consumption of the system. A common compromise is to select a medium grade filter even though this may not necessarily provide the quality of air required. One way of avoiding this compromise and reducing the pressure loss in higher grade filters, is to increase the surface area through which the contaminated air can flow. The list below shows how we increase the working area of filters as the grade increases to maintain an acceptable working resistance. Face area 600x600mm in all cases

Type	Area m²
low grade pad filter 50mm deep	0.36
medium grade 4 bag filter 400mm deep	1.92
high grade 6 bag filter 600mm deep	5-5.76

Modern developments include the 'rigid pack' paper filled filter a 600x600x300mm unit can provide a working area up to 18m²

Systems that concentrate movement of air also concentrate the dust and grime contained in it. Filtration is the only way to remove it.

FILTER REPLACEMENT

If filters are to carry out their role of dust extraction from the airstream whilst not affecting the air movement considerably they must be replaced by clean filters at regular intervals. This can be carried out using routine maintenance or condition based maintenance.

Routine maintenance involves making a decision based on previous experience, knowledge of dust conditions in the building and filter performance to determine a period after which the filter should be changed. So for example the filters in an air conditioning system may be routinely changed every six months. This is a simple method but may mean that the system operates with dirty filters for a time if the filters have clogged up quicker than expected. It could also mean that if the dust load is low relatively clean filters are being removed and replaced.

Condition based maintenance avoids the problems encountered with routine maintenance. Filter changing is based on the actual state of the filters rather than an assumption of their condition. The system works as shown in figure IP19. Transducers either side of the filter monitor air pressures. The pressure at point P_1 will always be higher than P_2 due to the resistance of the filter. However when the filter begins to clog up this pressure differential will increase. The pressure transducers can be observed manually or the signals fed into a BEMS system which will inform the building operators that the filters need changing.

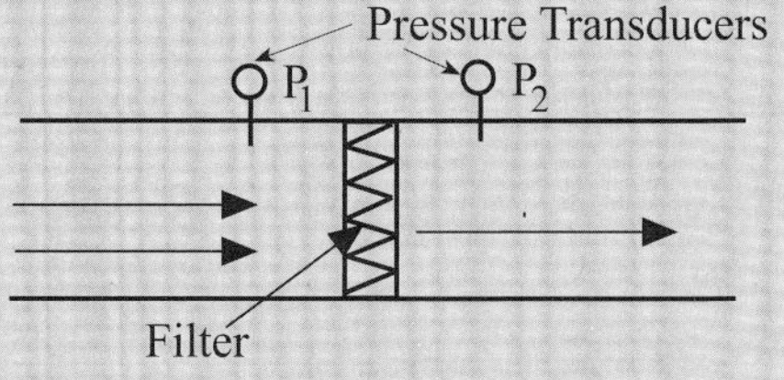

Figure IP19. Monitoring filter condition

of gaseous contaminants and odours. While most activated carbon filters are made of base carbon, the carbon can be impregnated to improve its ability to adsorb certain types of contaminant such as nitrogen dioxide(NO_2) and sulphur dioxide(SO_2) which are particularly damaging to documents and works of art. Carbon filters can be constructed of loose carbon, bonded carbon biscuits, carbon impregnated paper or fibre and pleated granular mat depending on the application. The filter is commonly housed in a metal frame and should always be preceeded by a pre-filter. In certain instances the carbon can be reactivated and reused at the end of its life.

The degree of effectiveness of a carbon filter is generally related to the amount of time that the air spends within the carbon. This is known as the dwell time. The greater the dwell time (lower the air speed or greater the carbon area) the more effective the carbon filter will be at odour or gas removal. Pressure losses through carbon filters can be high and manufacturers should be consulted to select a filter to optimise gas removal and minimise pressure drop.

7.2 Heater Coil

One of the functions of an air handling unit is to heat the incoming airstream. This can be achieved using direct heaters such as electrical heating elements. However it is more commonly achieved using heater coils. Heater coils are composed of a staggered grid of copper pipes conveying heated water between flow and return headers as shown in figure 7.5. The pipes can be connected by return bends which allows the flow and return headers to be at the same side of the heating coil. Coil heat output is improved by increasing the number of pipe rows. Low, medium or high temperature hot water or steam flows through these pipes in parallel assuring equal distribution of heat across the heater coil face. Attached to the surface of each tube are aluminium or copper fins. These fins increase the surface area for heat transfer between the hot coil and the airflow. Further increases in heat output are possible by corrugating the fins, but this does also increases the resistance the air experiences when passing through the coil.

Temperature control is achieved by fitting a temperature sensor into the room being heated or the extract duct work. This signal is fed into a control unit and is used to set the position of a valve supplying hot water from the boilers to the flow header. If the temperature of air in the extract duct work is higher than the room set point then the hot water flow to the coil will be modulated down or shut off. Thereby preventing further unnecessary heating of the room.

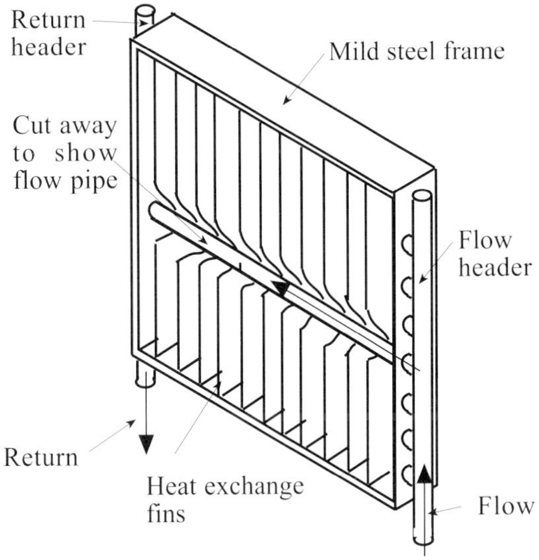

Figure 7.5 Heater coil

7.3 Cooling Coil

Cooling of the airstream is achieved by bringing it into contact with a cold surface. The cold surface is a cooling coil. The cooling coil can be either a direct expansion (DX) cooling coil which is the evaporator of a vapour compression chiller or it may be a water coil similar to the heating coil described above. It differs from the heating coil in that a mixture of water and antifreeze (glycol) circulates through it rather than hot water. This mixture is cooled using a chiller.

WASTE HEAT REJECTION

Vapour compression chillers have been described previously in section 5.1. An alternative method of cooling called absorption cooling is described in section 5.2.

IP20 - REFRIGERATION.PLANT.EFFICIENCY

Refrigeration plant is equipment which converts electricity into coolth. Coolth is the same stuff as heat i.e. thermal energy except that it has a negative value. In other words it is heat being removed from something causing a reduction in temperature. Most refrigeration systems are based on the vapour compression cycle and this information panel will concentrate on this system.

The energy consumption of a cooling system is by two principle components. Firstly by the compressor, and secondly by any fans or pumps used to remove waste heat from the condensor and remove coolth from the evaporator. The compressor consumes the majority, approximately 95%, of the energy input to the refrigeration equipment.

It is unfortunate that cooling is required in summer as this results in reduced system efficiency!

COEFFICIENT OF PERFORMANCE (COP)

The efficiency of a cooling system is normally called the coefficient of performance or COP. The COP of a real system is given by;

$$COP = \frac{\text{Cooling Capacity (kW)}}{\text{Total Power Input (kW)}}$$

Typical COP's for cooling are 1.5 to 2.0 which indicates that you can achieve 2kW of cooling for the consumption of 1kW of electricity. This compares with a COP of 2.5 to 3.0 if the refrigeration equipment is being used as a heat pump. To study factors affecting performance more clearly it is useful to look at the formula for the theoretical COP. This is given by;

$$COP = \frac{T_1}{T_1 - T_2}$$

where T_1 = Evaporator ambient air temperature (K)
T_2 = Condenser ambient air temperature (K)

From this it can be seen that for the COP to be high the difference between T_1 and T_2 should be small. Since T_1 is set by system requirements the variable is T_2 , the ambient temperature surrounding the condenser. taking a split cooling system as an example, what this means in practice is that when the outside air temperature increases the efficiency of the system will fall. This is unfortunate since most cooling is required in summer when high ambient condenser temperatures prevail. However, there are a number of things that can be done to improve this. The first is to increase the size of the air cooled condenser and ensure that there is good airflow through the device to keep the ambient temperature low. The next is to consider evaporative condensers (page 95) these use latent heat removal to reduce ambient temperatures. Finally, stable low temperature heat sinks should be considered such as surface and groundwaters.

ICE STORAGE

The efficiency of a chiller varies with the amount of work it is required to do. At low loads the efficiency will be reduced. One way of overcoming this problem is to use an ice storage system. Ice storage involves using an undersized chiller to produce an ice slurry during the night. The ice is stored in an insulated tank. During the day the chiller would not have enough capacity to satisfy the cooling demands of the building. However by operating the chiller at full load and drawing additional coolth from the ice store the building cooling demand can be satisfied.

The advantages of the system is that the chiller operates most of the time at full load and hence peak efficiency, it also provides a good proportion of the cooling requirement of the building using cheaper night time electricity tariffs. Both of these contribute to reducing operating costs.

This section will look at methods of condenser heat rejection.

Chillers generate a large amount of waste heat. In a domestic refrigerator, which we have used as an example of an every day chiller previously, the waste heat is simply allowed to enter the kitchen via the condenser coil at the rear of the refrigerator. However, in air conditioning systems the amounts of waste heat involved are too great and would cause serious overheating in the plant room. Because of this the waste heat must be safely rejected outside the building. There are three main ways in which waste heat is removed from the condenser. These are by using; air cooled condensers, evaporative condensers or water cooled condensers.

Air cooled condensers (figure 7.6) have been described
previously in their application to split air conditioning systems. They are predominantly used for smaller cooling loads (less than 100kW) although they are used for rejecting up to three times this value mainly due to the fact that water is not used in their operation. This means maintenance costs are low.

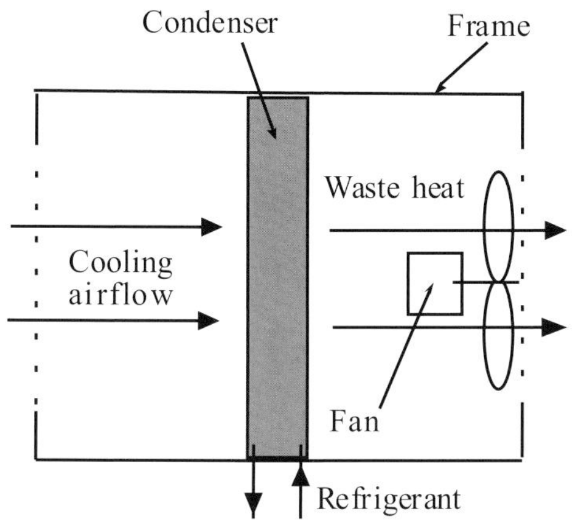

Figure 7.6 Air cooled condenser

Evaporative condensers. Are similar to air cooled condensers except that their heat rejection capacity is increased by spraying water over the condenser coil (figure 7.7). As this water evaporates it absorbs heat. Cooling of the condenser coil is therefore achieved by both sensible and latent means. The practical implication of this is that the unit has a smaller physical size for a given heat rejection capacity than an air cooled condenser.

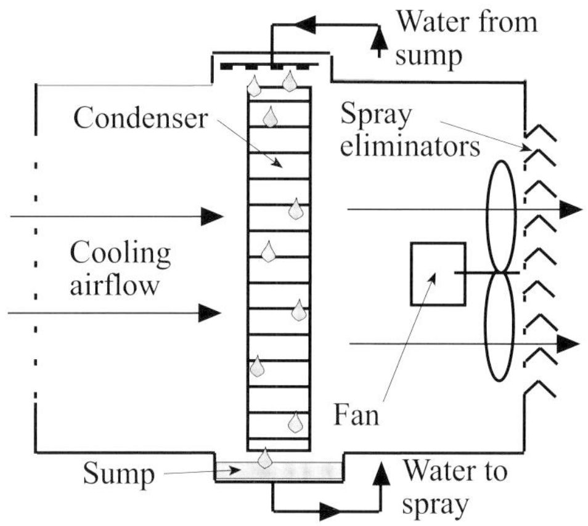

Figure 7.7 Evaporative condenser

The water circulating over the condenser is treated to prevent bacterial growth. In addition spray eliminators must be used to avoid water droplets, which may be contaminated, leaving the unit. Evaporative condensers can be used up to 500kW cooling capacity.

Water cooled condensers. Variations in ambient air temperature cause changes in the efficiency of air cooled condensers (see IP20). A more temperature stable heat sink is water. Water cooled condensers make use of this by jacketing the condenser in a shell which is filled with water (figure 7.8).

The condenser passes its waste heat to the water increasing its temperature by about 5°C. The water is then pumped to a water to water plate heat exchanger. Water from a large nearby source, such as a canal, river, lake or sea is also circulated through this heat exchanger having first been strained and filtered. In this way the condenser cooling water only makes thermal contact with the heat sink water such as canal water. The canal water having picked up heat from the condenser circuit is returned to the canal where the heat it carries is dispersed. The condenser cooling water leaves the plate heat exchanger and returns once more to the condenser to pick up more waste heat. The use of bodies of water such as rivers and canals as a heat sink is subject to water authority approval.

IP21 - PSYCHROMETRIC.CHART-STRUCTURE

The psychrometric chart shown in information panel 23 (page 108) looks daunting because of its complexity. However it is a very useful tool for studying the relationships between the temperature and moisture content of air. This information panel explains the structure of the psychrometric chart by breaking it down into simple components.

STRUCTURE

The psychrometric chart is like a sheet of graph paper. Instead of the normal x and y axese creating a square grid it has a number of axese and some of the grid lines are curves rather than straight lines.

Axis 1 is dry bulb temperature (DBT). This is like the x axis on a standard x-y graph. Any points drawn on the vertical lines on the graph will all have the same DBT. Dry bulb temperature is the temperature taken by a normal mercury in glass thermometer - units °C.

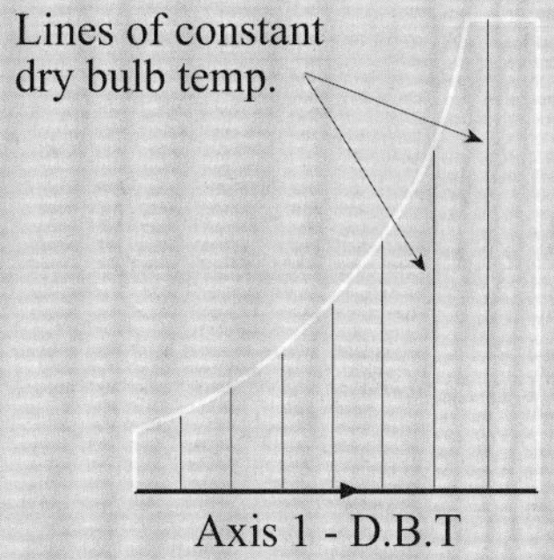

Axis 2 is moisture content. This is like the y axis of an x-y graph but is on the right hand side of the x axis. Any points drawn on the horizontal lines have the same

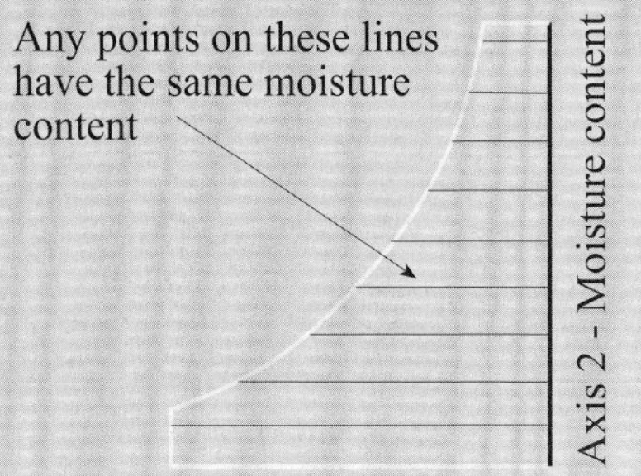

moisture content - units kg/kg i.e the weight of moisture (kg) in 1 kg of dry air.

Axis 3 is wet bulb temperature (wbt). As you can see this axis is curved and the lines of equal wbt are diagonal, sloping down from the axis. Wet bulb temperature is the temperature taken using a mercury in glass thermometer but with the bulb covered in damp cloth - units °C. Evaporation of water from this cloth cools it down so the wbt is usually lower than dbt. The amount of evaporation and hence temperature depression depends on the moisture content of the air. The dryer the air the greater the evaporation and the greater the depresion of wbt below dbt.

Axis 4 is relative humidity (RH). This is a short y type axis on the left hand side of the dbt axis. As you can see the lines of equal RH curve upwards from this axis. One point to note is that the 100% RH curve forms the wbt axis. This curve is also known as the saturation curve.

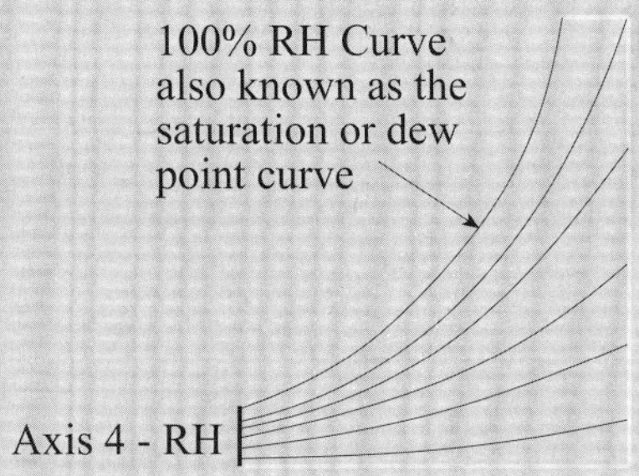

By overlaying each of the axese and related constant lines the psychrometric chart is formed. (see page 102 for its use and page 108 for the chart)

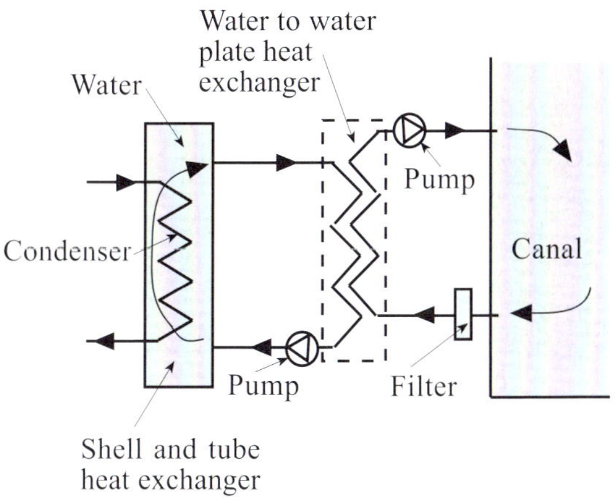

Figure 7.8 Water cooled condenser

Cooling Towers. In locations where there are no large bodies of water that can be used as a heat sink, the water cooled condenser is used in conjunction with a cooling tower. A cooling tower is a device which cools the condenser cooling water by evaporation before returning it to the condenser to collect more heat. Figure 7.9 shows a forced draught cooling tower. It can be seen that the condenser cooling water is allowed to tumble down through the device whilst air is forced upwards through the cascading water by a fan.

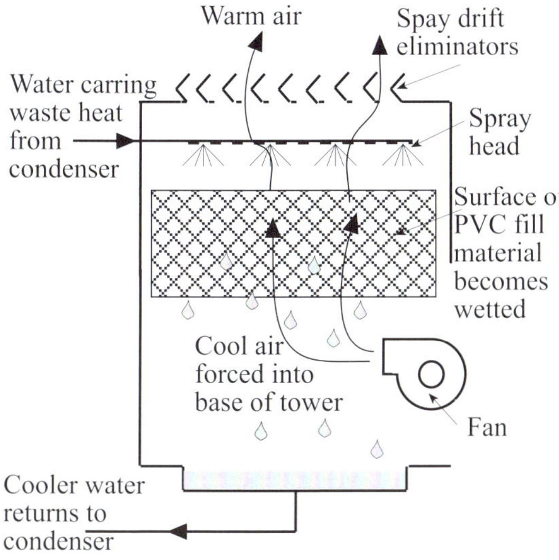

Figure 7.9 Forced draught cooling tower

The purpose of the tower is to enhance evaporative cooling of the condenser water. It does this by increasing the surface area of the water exposed to air. Evaporation is a surface effect so increasing the surface area of water in contact with the air increases evaporation. Water surface area increases are achieved in a number of ways such as by allowing the water to tumble down splash bars, by spraying or by running it over a PVC matrix. When the condenser water evaporates it absorbs latent heat from the water which is left behind. The effect of this is to cool the water which collects in the sump at the base of the tower. This is pumped back to the water cooled condenser to remove more waste heat.

Cooling tower hygiene is an important area of concern since the water temperatures in the tower are conducive to bacterial and algal growth. In particular legionnaires disease, which is a form of pneumonia, has been associated with wet heat rejection equipment. The legionella bacteria grow in the warm water of the cooling tower. They escape from the tower as part of the mist created by the flow of air and water through the tower. If ambient conditions are suitable and the bacteria carrying droplets are breathed in by a susceptible passer by a potentially fatal infection can occur. The problem is avoided by using air cooled condensers. However, dry heat rejection uses approximately 30% more energy for the same capacity as a wet method. Cooling towers can be used safely with the following precautions;

- Use of spray eliminators to prevent the release of infected droplets

- The tower should be built of easily cleanable materials such as plastics or epoxy coatings with smooth surfaces. Access doors should be incorporated into the tower to facilitate cleaning.

- The tower should be positioned away from air intakes which could draw infected droplets into the building through the air conditioning system.

- A programme of maintenance and cleaning should be carried out throughout the life of the tower. This should include dosing the cooling water with bactericides. Chemicals to prevent algal growth should also be used since algae tend to coat surfaces and give the legionella a medium on which to grow.

7.4 Humidifiers

The amount of moisture in a given volume of air is most often stated in terms of its relative humidity (RH). This is a measure of how much water vapour there is in the air sample compared to its saturated state. Completely dry air would have a relative humidity of 0% . Air which is saturated would have a relative humidity of 100%. For human comfort the relative humidity of the air in a room should be between 40 and 70% RH. If the air is below 40% RH the air will feel dry and lead to discomfort through dry eyes and throats. It is also known that the risk of static shocks and problems with VDU screens increases in dry atmospheres. Relative humidities above 70% result in discomfort due to clamminess and overheating. This is because the body's normal mechanism for cooling itself down, sweating, cannot operate effectively in a humid environment. Prolonged relative humidities above 80% can lead to mould growth in buildings.

In addition to human comfort, some industries require stable relative humidities for the production and storage of materials without degradation. Examples are the high relative humidities required in the textile industry, typically 65% in wool processing and 75% RH in cottons, to avoid problems such as electrostatic build up and yarns breaking 50-55%RH is required in the print industry to prevent sheet papers curling and breaks in newspaper webs. At the other end of the scale, low relative humidities are required by some industries such as in car panel manufacture to avoid corrosion.

Low relative humidities occur when cold outside air is brought into the building and is heated. For example the relative humidity of outside air at 0°C and 90%RH drops to 23%RH when heated to 20°C. The problems associated with this can be overcome by adding moisture to the airstream (humidifying it). High relative humidities occur when warm summertime air is cooled or in spaces with open bodies of water such as swimming pools. Problems associated with high relative humidities can be avoided by removing moisture from the airstream (dehumidifying it). Dehumidification is discussed in the next section.

This section will discuss methods for humidifying a space. Humidification systems are categorised by the way they deliver water vapour to the air in a room. The two categories are direct and indirect humidification.

Direct humidification is used in industrial situations and involves adding moisture directly into the air of the room in which humidification is required.
Indirect humidification is used in buildings with central air conditioning systems. The air is humidified within the air handling unit and is then delivered to the room using ducting.

There are two general methods of humidification. These are; wet humidification and steam humidification.

WET HUMIDIFIERS

Wet humidifiers work by encouraging liquid water to evaporate. This creates water vapour which mixes with the airstream to humidify it. For the water to evaporate it must absorb heat from its surroundings. As a result wet humidifiers cause the airstream temperature to fall during the humidification process. To overcome this problem, in air handling units, a pre heater initially warms the incoming airstream. The warmed air then passes through the humidifier but becomes cooled in the humidification process. The air must then pass through a reheat coil to bring the airstream up to the required temperature.

Wet humidifiers can take a number of different forms. The common feature of each is that they all aim to increase the surface area of water over which evaporation can take place.

Air washers are used mostly in industrial humidification. As the name implies they provide the dual function of humidifying the airstream and at the same time washing out some dust and odours. The airstream is made to flow smoothly by passing between baffle plates (figure 7.10), it then passes through a fine mist of water droplets created by a spray head. This provides the contact between the liquid water and the air necessary for evaporation to take place. Spray eliminators are placed downstream from the humidifier to prevent the carriage of liquid water further down the ducting.

Evaporation of the water cools the airstream, if this is desirable, further cooling can be obtained by using a

NOTES

chilled water spray. During periods of warm weather the water in the sump may remain still for a long period. It is important that this water is treated to avoid bacteriological growth which could lead to infection of the building occupants. For example, Humidifier fever, an industrial disease with flue like symptoms, is commonly associated with humidifiers with reservoirs. However, care should be taken in the choice of biocide and that it is not carried by the airstream into the working environment.

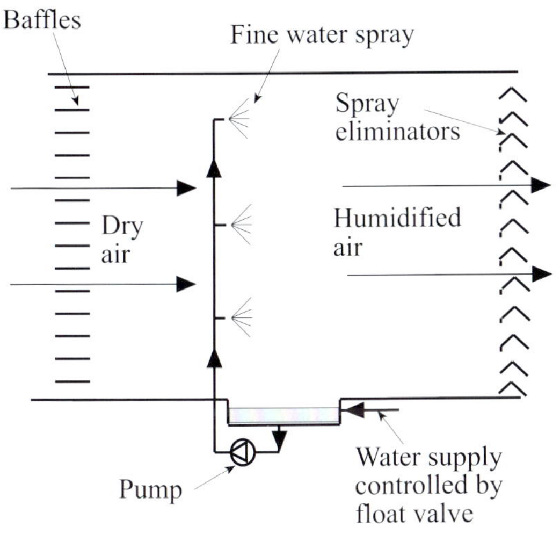

Figure 7.10 Air washer

Capillary washers are humidifiers with a better humidification effectiveness than the basic spray air washer. The greater effectiveness is obtained by directing the spray onto a matrix of metal, glass or plastic fibre cells (figure 7.11). The spray coats these cells resulting in further spreading out of the water due to capillary action. The airstream passes between the gaps in the cells and hence over the wetted surfaces. In this way close contact is obtained between the airstream and extended water surface. The airstream can be in the same direction as the spray or in the opposite direction to the spray, termed parallel and contra flow respectively.

Ultrasonic humidifiers create an extremely fine mist of water droplets by passing liquid water over a ceramic plate which is made to vibrate at ultrasonic frequencies using the piezoelectric effect, (figure 7.12). The small droplet size results in quick and effective evaporation

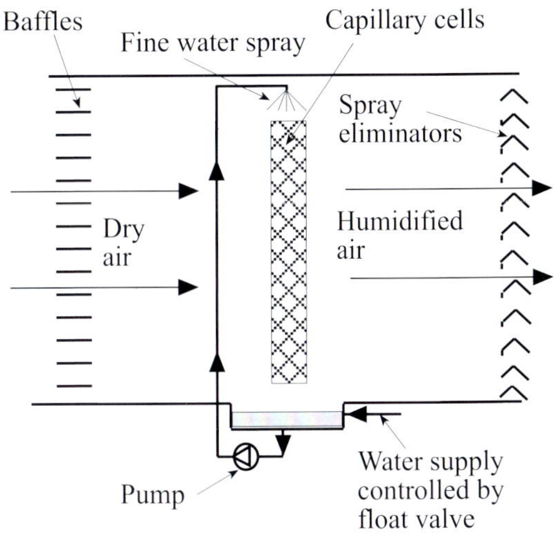

Figure 7.11 Capillary washer

of the water. The consequence is a rapid adjustment in relative humidity in response to a call for humidification from the controls. The excellent evaporation characteristics of the device make it very efficient in energy terms and ultrasonic humidifiers have an extremely low energy consumption for a given humidification load.

However, ultrasonic humidifiers must be supplied with demineralised water. This is to avoid scale build up which would otherwise lead to clogging.

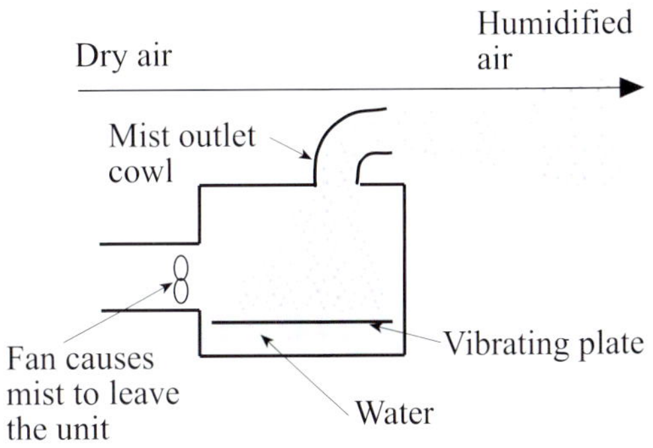

Figure 7.12 Ultrasonic humidifier

Atomising nozzle humidifiers produce a fine spray of cold water directly in the air or within air handling units. Droplet size is small so evaporation and hence

IP22 - PSYCHROMETRIC.CHART-USES

In information panel 21 the structure of the psychrometric chart was explained. The psychrometric chart is used to illustrate the relationships between the temperature and mosture content of air. The first of these is to show, graphically, the condition of air in a room. This is achieved by placing a point on the chart representing the state of the air. This point appears where the variables intersect. It can also be used to visualise the changes which take place following heating, cooling or humidification.This information panel gives an example of each of these uses. Firstly to determine a full set of variables describing the current condition of the air in a room and secondly to track changes in moisture content,temperature and relative humidity.

DETERMINING AIR CONDITIONS

The current condition of the air in a room can be illustrated on the chart if two of the four variables wet bulb temperature, dry bulb temperature, relative humidity or moisture content (wbt, dbt, RH or mc) are known. For example assume the wbt and dbt of outside air are 4°C and 5°C respectively. These values can be used as two coordinates to create a point (A) on the chart (see IP23). Having got this point the chart allows us to determine the other two variables i.e. RH and moisture content. They are found by tracking the point back to the relevant axis along the lines of constant value. In this case they are 85%RH and 0.0046kg/kg (4.6g/kg) respectively.

FOLLOWING CHANGES TO CONDITIONS

For illustration we can assume that the air described by point A on the psychrometric chart is outside air in winter.

A to B - Heating

If this air is used to ventilate a building either through infiltration or mechanical means it will come into contact with a heating appliance such as a radiator. This will raise its temperature to 21°C. As a result the condition of the air changes. This is illustrated by a move from point A to B on the chart. The most striking change is the fall in relative humidity from 85 to 29% RH. Note the relative humidity has changed but the moisture content is has not. It remains at 4.6 g/kg.

B to C - Addition of moisture

If the heated room air described by point B comes into contact with a source of water vapour such as human breath, a kettle boiling or a humidifier it will absorb moisture. If we assume 5.4 grammes of water vapour per kilogram of dry air are added, then the condition of the air changes from point B to C on the chart. Three variables have changed these are;wbt to 16.8°C, mc to 10.0g/kg and RH to 64%, the dbt stays the same at 21°C as no heat has been added or removed from the air.

C to D cooling

Now imagine that the air is cooled in some way. It may for example be cooled by an air conditioning unit or it may simply come into contact with a cold window. If the air is cooled by 7°C to 14°C then the condition of the air moves from point C to D. The most noticable change is that the RH has increased to 100%. The air is said to be saturated. The values of dbt and wbt are the same. The temperature at which this occurs is known as the dew point temperature as it is the temperature at which condensation is just starting to occur.

D to E Further Cooling and Condensation

If the air is cooled by a further 4°C the air condition cannot leave the boundaries of the chart and so runs down the saturation curve to point E. In practice this means that 2.5 grams of moisture per kilogram of dry air will be condensing on the cold surface. This is how condensation occurs on cold windows and is the basis of moisture removal from the air by heat pump dehumidification.

The structure of the chart is shown on page 96, the chart itself and the above examples are on page 108.

humidification is rapid. Water is supplied directly from the mains, avoiding any contamination risk, and compressed air is used to create the spray. The small size of the atomising nozzle means that it can easily be blocked by mineral build up. To avoid this a needle built within the nozzle head periodically cleans the orifice automatically. When used in air handling units all of the spray released from the nozzle evaporates removing the need for water recirculation and chemical treatments. Atomising nozzle humidifiers provide close control of humidity at low running cost and with low maintenance requirements.

STEAM HUMIDIFIERS

Unlike wet humidifiers, steam humidifiers do not chill the airstream during the humidification process. This is because the moisture is delivered to the airstream already in the vapour state (as steam) having been created by a heating element.

Electrode-boiler humidifiers (figure 7.13) are the most widely used type of steam humidifier in direct and indirect humidification due to their low cost and ease of installation.

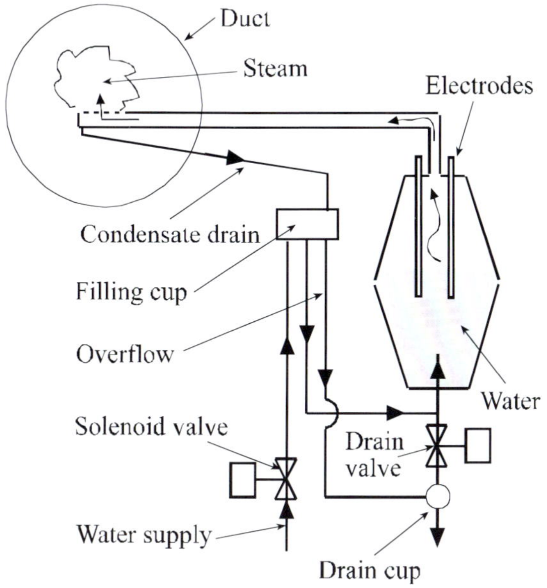

Figure 7.13 Electrode-boiler humidifier

The core element is a small boiler comprised of a non-conductive polypropylene tank fitted with three or more bare steel electrodes. When the tank is filled with water the electrodes become immersed. Electrical connections are made to the electrodes and current flows directly through the water causing it to heat up and boil. Output of the unit is controlled by varying the depth of water in the tank. Continual boiling of the water causes the concentration of minerals in the tank to build up. To avoid this there is an automatic cycle of emptying and re-filling the tank with fresh water. When the boiler eventually scales up it is simply replaced or opened and de scaled. The primary disadvantage of this system is high running costs and the need to regularly replace boilers.

Resistive element humidifiers (figure 7.14) are like small kettles boiling the water within them using an electric element. Regular drain and refill cycles prevent excessive scale build up. Switching off individual elements and modulating the power supply provides very close control of steam output, making them the preferred choice for close control applications.

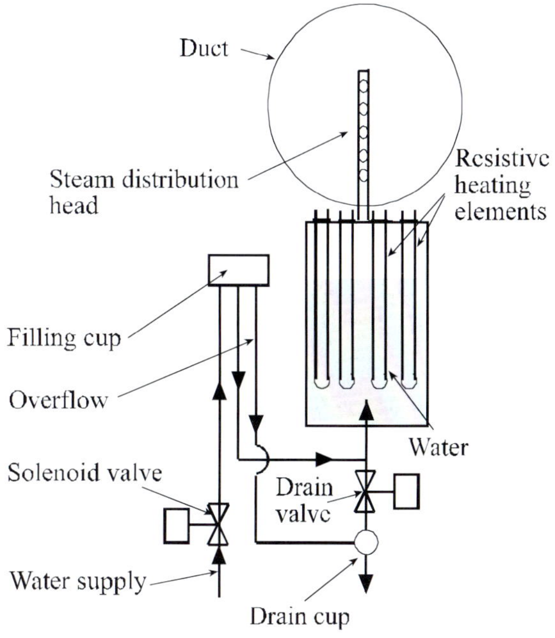

Figure 7.14 Resistive element humidifier

Gas-fired steam humidifiers use a gas heater to boil water and create steam. Gas is approximately four times

calorex®
HEAT PUMPS & DEHUMIDIFIERS
Leaders in our field
Specialist designers and manufacturers of dehumidifiers and pool comfort control systems
Commercial Dehumidification
Process Drying by Dehumidification
Swimming Pool Dehumidification & Heat Recovery
Specialist Low Humidity Control
Calorex Heat Pumps Ltd
The Causeway, Maldon, CM9 5PU
Tel: 01621 856611 Fax: 01621 850871
Email: sales@calorex.com Web: http://www.calorex.com

cheaper per unit of energy than electricity. As a consequence the running costs of gas fired steam humidifiers are low, making them increasingly popular for humidification. Their structure is similar to the gas boilers described in section 1.2 except that the water is heated to boiling point by the burner.

7.5 Dehumidifiers

The relative humidity of a sample of air can be reduced using two principle techniques; cooling below the dew point temperature and chemical adsorption.

Dew point dehumidification. Humid air contains invisible water vapour. This only becomes visible when it changes back to liquid water. This phase change is achieved by cooling the humid air until condensation starts to occur. This is what happens when air in a room touches a cold window. It becomes chilled and condensation forms on the cold glass surface. However, this only happens if the glass is cold enough. The threshold temperature below which condensation occurs is called the dew point temperature (see IP22).

Dehumidification of air occurs in the same way. The air is made to pass over a cold coil in the air handling unit which is below the dew point temperature of the air. This causes some of the water vapour in the air to condense out onto the coil where it is drained away. In some dehumidification applications the condensate can be collected and re used. One example is in swimming pool dehumidification where the condensate is used to top up the pool to offset the use of some of the mains water which must be purchased.

Dehumidification by chilling is an energy intensive process since the air must be reheated to bring it back up to comfort temperatures. One way of achieving this efficiently is to use heat pump dehumidification. Both coils of the heat pump are placed in the ducting as shown in figure 7.15. The first coil the air meets is the evaporator coil. This is cold and removes water from the air by condensation. The air then passes over the condenser coil of the heat pump which re heats the air using energy which, in a simple cooling situation, would go to waste. Both sensible and latent heat are recovered in this process. This is discussed more fully in IP13 (page72). For manufacturer see page 104.

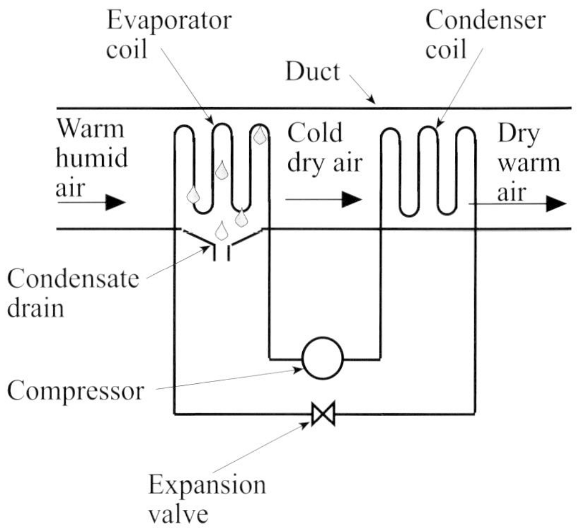

Figure 7.15 Heat pump dehumidification

Desiccant dehumidification involves the removal of water vapour from the air by chemical adsorption. The humid airstream (figure 7.16) is passed over a surface which is coated with a desiccant chemical such as silica gel. This removes water vapour from the airstream. The gel would quickly become saturated and unable to remove further water from the airstream. It must, therefore, be reactivated by heating. The desiccant chemical coats the tubes of a desiccant wheel. The lower part of this wheel absorbs moisture out of the airstream.

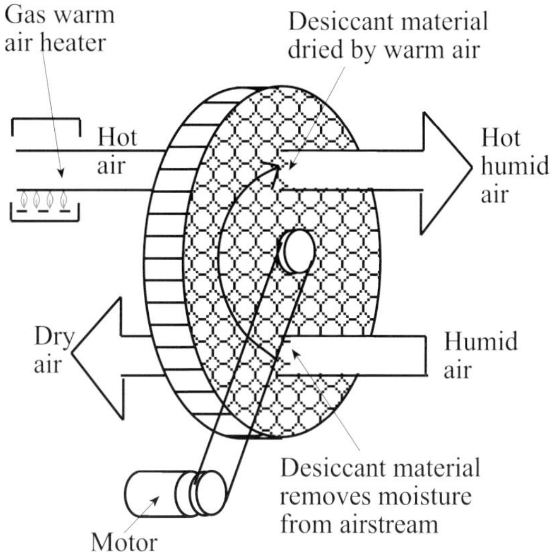

Figure 7.16 Desiccant dehumidifier

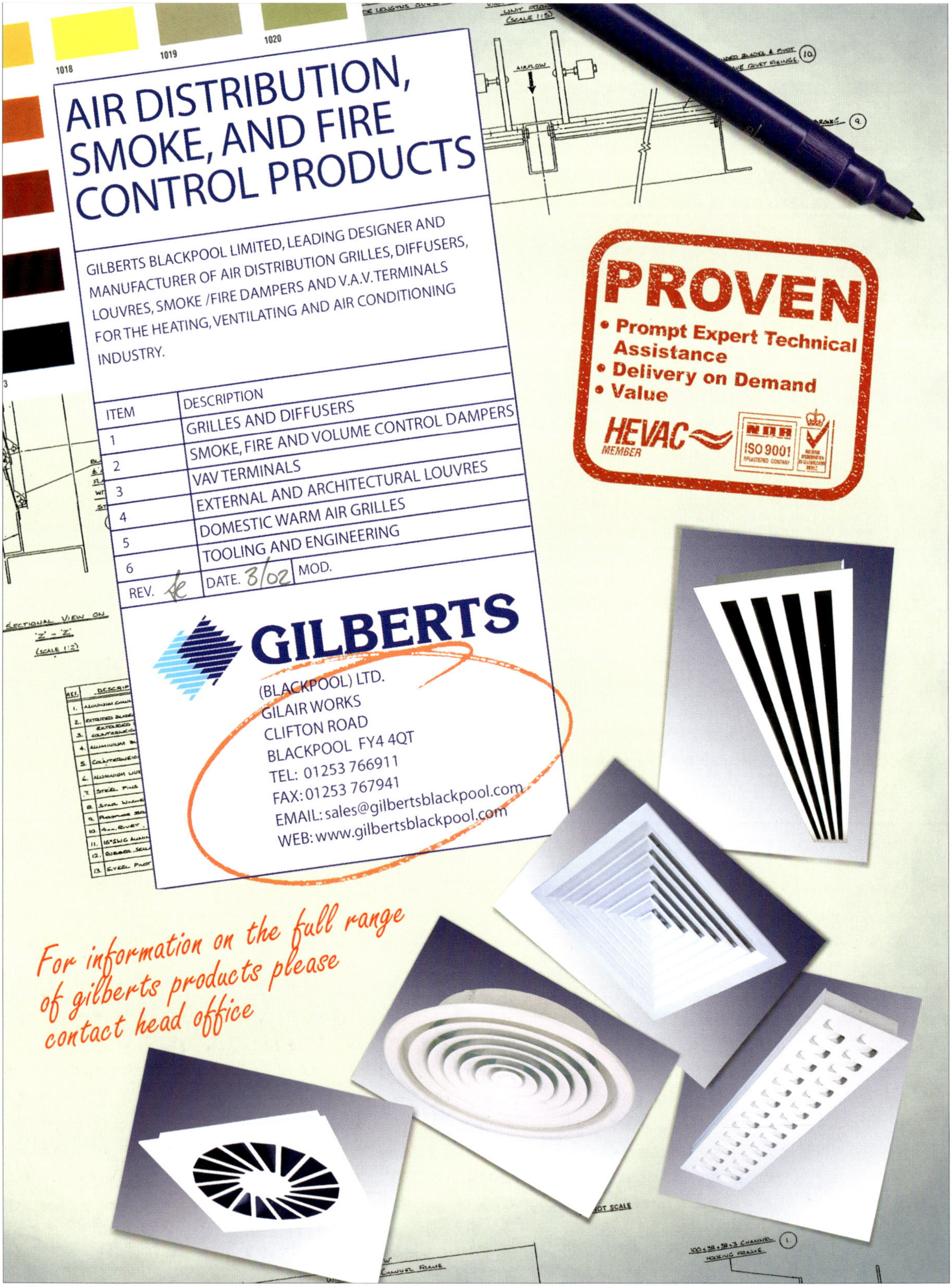
AIR DISTRIBUTION, SMOKE, AND FIRE CONTROL PRODUCTS
GILBERTS BLACKPOOL LIMITED, LEADING DESIGNER AND MANUFACTURER OF AIR DISTRIBUTION GRILLES, DIFFUSERS, LOUVRES, SMOKE /FIRE DAMPERS AND V.A.V. TERMINALS FOR THE HEATING, VENTILATING AND AIR CONDITIONING INDUSTRY.
ITEM
DESCRIPTION
1 GRILLES AND DIFFUSERS
2 SMOKE, FIRE AND VOLUME CONTROL DAMPERS
3 VAV TERMINALS
4 EXTERNAL AND ARCHITECTURAL LOUVRES
5 DOMESTIC WARM AIR GRILLES
6 TOOLING AND ENGINEERING
REV.
DATE. 3/02
MOD.
GILBERTS
(BLACKPOOL) LTD.
GILAIR WORKS
CLIFTON ROAD
BLACKPOOL FY4 4QT
TEL: 01253 766911
FAX: 01253 767941
EMAIL: sales@gilbertsblackpool.com
WEB: www.gilbertsblackpool.com
PROVEN
• Prompt Expert Technical Assistance
• Delivery on Demand
• Value
HEVAC MEMBER
ISO 9001
For information on the full range of gilberts products please contact head office

This section then rotates into a new section of ducting where warm air drives off the moisture and re generates the wheel. The now dry section of wheel rotates back into the humid airstream to continue the drying process.

7.6 Diffusers

One of the most important aspects of air conditioning systems is the effective input of conditioned air into the room in which it is required. The following criteria must be satisfied;

- The air should enter quietly so that it does not create annoyance with respect to the ambient sound environment present in the room.
- The air should achieve effective distribution to all parts of the room and achieve adequate mixing so that no stagnant zones exist.
- The air should enter the room without directly impinging on the room occupants. This would cause uncomfortable physical and thermal sensations (draughts)

These three criteria can be achieved by the correct selection and positioning of room air diffusers. There are many forms of room air diffuser. Some of which are shown in figure 7.17. The differences provide a choice of air distribution pattern and flexibility for accommodating different applications. The following section describes some of the more common diffusers and uses a square ceiling diffuser to explain basic airflow concepts. Manufacturers (page 106) should be consulted for recommended arrangement and spacing of diffusers.

Square ceiling diffusers form part of a suspended ceiling system. To evenly distribute the air across the room they are positioned at the centres of a 3 to 4m square grid covering the space. Figure 7.18 shows how air enters the diffuser horizontally but is then deflected through 90°, heading straight down towards the floor. This situation would be unacceptable since the airstream would hit any occupants below the diffuser. Draughts are avoided by using vanes within the unit to guide the airstream horizontally along the ceiling. This keeps the airstream out of the space in which the people work, known as the occupied zone. The occupied zone is considered to be any space in which occupants linger for a "significant" time. Physically it is a volume within the room with a height of 1.8m (comparable to a typical occupant) and bounded by a perimeter 0.15m from the walls.

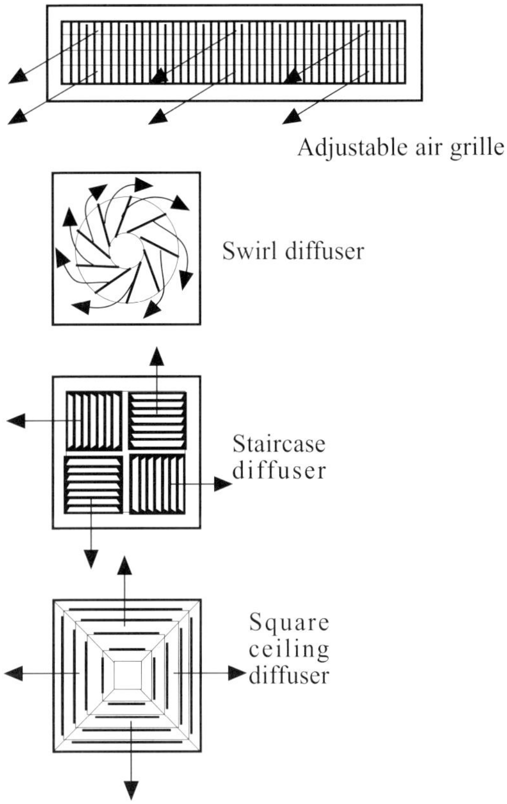

Figure 7.17 Four common types of diffuser

The airstream on leaving the diffuser moves along the underside of the suspended ceiling mixing with the room air as it does so by a process known as entrainment. The distance the airstream moves from the diffuser whilst maintaining a speed over 0.5m/s is known as its throw. The throw length is increased because the air leaving the diffuser experiences friction with the suspended ceiling. The result of this, known as the coanda effect, is to hold the airstream next to the ceiling. The distance covered before the airstream drops into the occupied zone is therefore increased. The texture of the underside of the ceiling and the presence of any projections can disrupt the throw of the diffuser.

IP23 - PSYCHROMETRIC.CHART-DIAGRAM

Reproduced by permission of the Chartered Institution of Building Services Engineers. Pads of charts for calculation and record purposes are available from CIBSE, 222 Balham High Road, London, SW12 9BS, UK

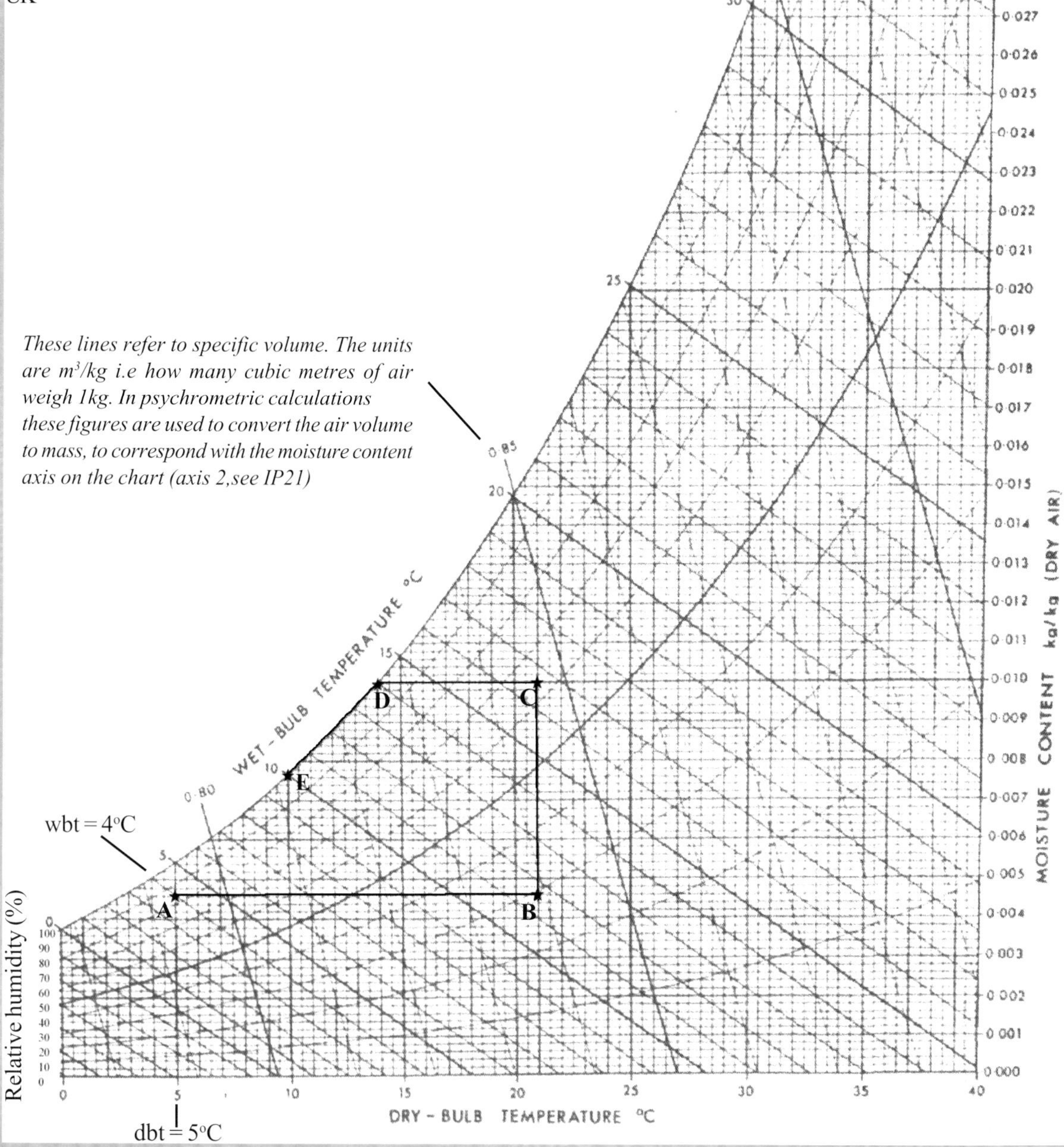

This is a shortened version of the psychrometric chart the actual CIBSE chart runs from -10 to 60°C dbt and gives information on enthalpy and specific volume.The structure of the chart is shown on page 96 and example uses described on page 102

For example use of surface mounted luminaires would cause the airstream to be deflected downward into the room.

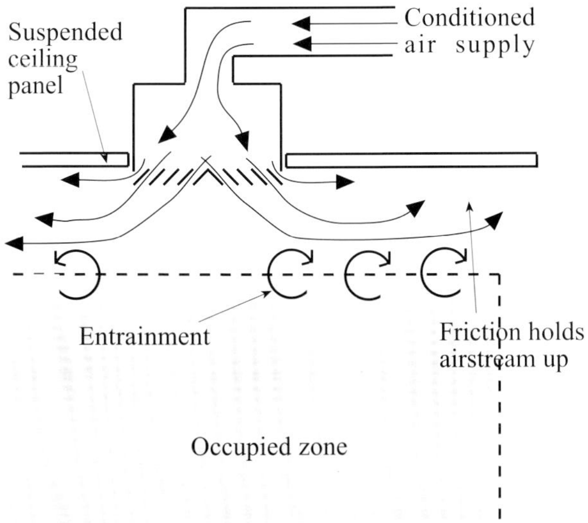

Figure 7.18 Suspended ceiling diffuser

When the airstream drops into the occupied zone its velocity should be no more than 0.25m/s in cooling/ summer mode or 0.15 m/s in winter. The former higher airspeed is considered acceptable in summer when its cooling effect is advantageous. However in winter this high airspeed would be felt as a draught so a lower speed is specified.

The airflow pattern from square suspended ceiling diffusers is in four directions perpendicular to each other. The airflow in any of the four output direction can be modified using additional adjustable vanes and dampers within the diffuser.

As well as being ceiling mounted, diffusers can also be mounted in walls or floors. A simple wall mounted diffuser located just below the ceiling would send a stream of air parallel to the ceiling into the room. The coanda effect would once again assist throw, air entrainment and avoidance of inappropriate air entering the occupied zone.

The operation of floor mounted diffusers requires careful consideration since they are within the occupied zone. The air velocity leaving the diffuser must be low, it is also beneficial if it is made to swirl. Both of these reduce the risk of discomfort as air enters the room. Adequacy of mixing with the existing room air is assisted by movement of the occupants and convection currents set up by temperature differences between the input supply air temperature and the temperature of the occupants and office equipment. The ultimate manifestation of this is displacement ventilation which is discussed in section 8.0.

It is not always possible to mount an array of diffusers across a space which is tall and has no suspended ceiling, although some installations do leave an array of ducting and diffusers exposed as a feature. The alternative is to use jet diffusers (figure 7.19). These devices produce a jet of air with a very long throw which is suitable for supplying air to large atria, halls, factories and leisure complexes such as swimming pools. They are constructed as a ball and socket and so are highly adjustable in terms of direction of throw.

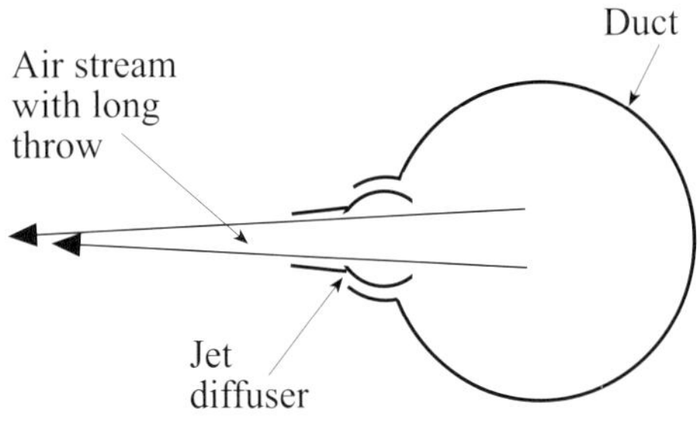

Figure 7.19 Jet diffuser

POSITIONING OF SUPPLY DIFFUSERS

Manufacturers will supply information on the best positioning of diffusers to obtain optimum performance. This information is available for individual diffusers based on experience obtained from actual installations and also laboratory testing using test rigs and room mock-ups. It has been found that the performance of diffusers depends on;

- dimensions of the room - any air stream entering a room horizontally say from within the ceiling will travel along it until it meets a projection or a vertical wall. When it does so it will be deflected downwards.

NOTES

www.info4study.com

- texture of surfaces - The distance an airstream in contact with a surface travels depends on the friction experienced between it and the surface. Highly textured surfaces will increase friction and reduce throw.

- ambient temperatures - Strong sources of heat in a room such as office equipment will set up a strong upward convection current. These could disrupt design diffuser/room airflow patterns and should be considered if possible during the design.

- air velocities - the velocity of air entering a room through a diffuser has a major influence on the travel of the airstream. However the speed of air movement also effects the amount of noise produced at the diffuser. As a result in quiet office conditions air velocities must be limited to reduce noise levels. In noisy locations such as leisure complexes or atria velocities can be higher

- whether the system will be predominantly used for heating or cooling. - chilled air entered at high level into a room would descend downwards. Heated air would tend to remain at high level due to its buoyancy. The opposite effect would occur with low level entry of chilled or heated air. The diffusers should be positioned to take advantage of these natural air movements to assist room air diffusion.

- Spacing of diffusers - diffusers should be suitably spaced so that airstreams from adjacent diffusers do not interact. For example if two airstreams running along the underside of a ceiling were to hit head on the tendency would be to cause a downward current midway between the two diffusers.

Following careful design, fine tuning of the system can take place on site using the adjustment available within the system provided by integral dampers and guide vanes.

EXTRACT GRILLES

Systems incorporating a return air path require an extraction grille. The locations of these are less critical than supply diffusers since the air flowing into them is at room temperature and is at a lower velocity than the incoming air. It is therefore less likely to cause an uncomfortable thermal or physical sensation if this air passes by an occupant. However the extract grilles do work as part of the system and their positioning can be used to improve the effectiveness of the ventilation system. The following factors should be considered;

- Incoming air will be filtered and cleaned to avoid staining of surfaces near the supply diffuser. However, extract air will be carrying dust from the room this, over time, will cause some staining of finishes near the extract grille.

- If the location of sources of pollution such as photocopier are known, then the exhaust grille should be positioned near to them. This will remove the pollutants from the room as they are produced. This also applies to heat pollution. A number of manufacturers produce extract luminaires which remove the heat given out by the lighting at source along with room extract air. Care must be taken if the ceiling void is used as an exhaust air plenum as the temperature of the suspended ceiling may increase. The effect of this is to increase thermal discomfort since the ceiling will act as a warm radiator. This is a particular problem in rooms with low ceiling heights.

- The exhaust terminal should not be close to a supply diffuser as this would short circuit the system causing supply air to leave the room without having had the opportunity to mix with the existing room air.

- For the ventilation system to work effectively it is important that all parts of the room benefit from the conditioned air entering it. If any stagnant zones exist such as in alcoves the extract grille could be positioned there to encourage air movement through these still areas.

7.7 Ducting

Ducting forms the distribution network for air based air conditioning systems. Their function is analogous to pipes in wet heating systems. However because of ventilation requirements and the low heat/cooling carrying capacity of air, ducting tends to have a relatively large cross sectional area. Because of this careful consideration must be made early in the building design stage to accommodate and integrate ducting runs into the structure and fabric of the building. This is espe-

cially so near the plant room where duct sizes are at their greatest.

Careful design of ducting systems is important as the way the ducting delivers air to the diffusers strongly influences the way air enters the room. The air should flow in a smooth manner as turbulence can change the air distribution characteristics of the diffusers.

The layout of the ducting dictates how much fan energy is needed to overcome resistance to airflow. Changes in the ducting such as 90 degree bends, size reductions and other components increase airflow resistance. Ducting runs should therefore be as simple and linear as possible (see design software page 114). The fan must be sized to overcome air resistance in the ductwork and so duct design has important implications for fan size and energy use. Ducts carry heated or cooled air. Loss of this conditioned air from leaks in the distribution network results in wasted energy and lack of control. This also results from heat transfers through the walls of the ducts. To overcome this ducting joins should be well sealed and ducts insulated where temperature differentials between air in the ducting and ambient conditions dictate it.

Finally, ducting can be a source of noise in the rooms being served. The sources are transfer of noise and vibrations from the plant room and noise due to airflow in the ducts. A range of anti vibration mountings and acoustic isolation joints are available to prevent vibration transfer. Sound attenuators can be positioned in the ducting to absorb airborne noise. Air flowing through ducting and dampers creates noise in itself. To overcome this there are restrictions on maximum air velocities in ducting and dampers should be positioned as far away from the room air outlet as possible.

7.8 Dampers

Dampers are analogous to valves in a water distribution system. They vary the volume of air flowing through ducting by restricting or extending the open cross sectional area of the duct. They are constructed from an array of blades which rotate about their central axis like a louvred window system. The rotation is driven by motorised actuators (see section 1.6) in response to signals from the building energy management system. Accurate and reproducible positioning of the damper blades by the actuator is essential for close control of the air delivery system (page 112).

Dampers can be grouped under the following categories;

Butterfly dampers are the simplest form of damper, they are composed of a single blade which can be positioned either parallel or perpendicular to the airflow (figure 7.20). These two positions give maximum or minimum airflows respectively. Their small size means they are widely used in terminal units or small branch ducts. A disadvantage of butterfly dampers is that they create a very turbulent airflow which can result in noise within the ducting system.

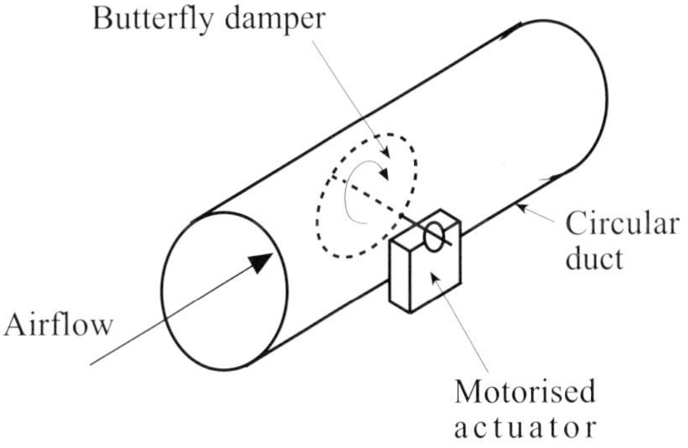

Figure 7.20 Butterfly damper

Multi blade dampers come in two patterns; opposed blade dampers (figure 7.21) and parallel blade dampers (figure 7.22).

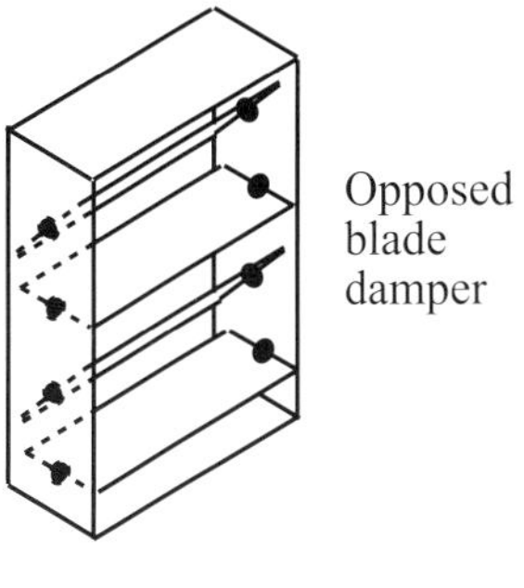

Figure 7.21 Opposed blade damper

NOTES

These dampers regulate the air flow rate through themselves by rotating the blades. Ideally there will be a linear relationship between blade angle and airflow however, in practice linearity is seldom achieved. Operation of opposed blade dampers restricts the airflow but does not affect the airflow direction.

Parallel blade dampers do change the direction of the airflow since the blades are all rotated in the same direction. This effect is used to control the direction of airstreams leaving a diffuser. It can also be used to mix fresh and recirculated air within the air handling unit using two duct mounted dampers.

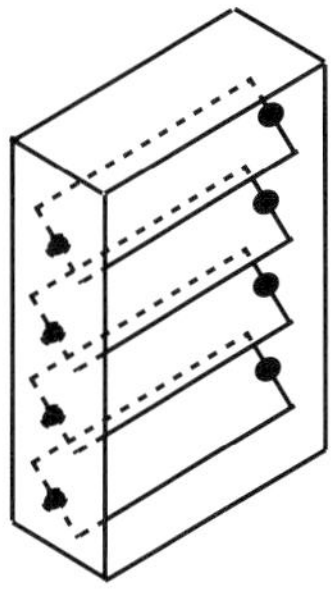

Figure 7.22 Parallel blade damper

FIRE DAMPERS

Fire and smoke can be spread through ducting from the source of a fire to other rooms unless it is prevented by the closure of fire dampers. Dampers can be made to close automatically in the event of a fire. The signal for the closure arises from the fire detection system. This may be linked into the building energy management system which can also shut down ventilation fans and/or operate smoke clearance fans.

Fire dampers can also be constructed from a single blade which falls into position across the duct, it does this when a fusible link burns out due to the high temperatures experienced in a fire. Finally, air transfer grilles painted with intumescent paint can be positioned across ducting. During a fire the intumescent paint expands to many times its normal volume and so closes the free area of the grille through which fire or smoke could spread.

7.9 Delivery Systems

Many modern office buildings with large glazed areas and heavy use of computers require year round cooling. However, the cooling load is unlikely to be the same in all parts of the building. Depending on equipment densities and orientation the cooling requirement will vary from area to area. Solar gains into south and west facing zones are a particular problem since they add considerably to the cooling loads caused by occupants, lighting and office equipment. There are three methods used by centralised air conditioning systems to heat or cool a space. The particular method used depends on the variation in heating/cooling load.

- If a large zone such as an open plan office requires cooling throughout the space then a single zone system can be used.

- Where the demand for cooling differs between spaces, control of individual temperatures is achieved by varying the amount of cooled air allowed to enter the room. This is achieved using a variable air volume (VAV) system.

- Where there is a need to control temperatures and at the same time maintain supply air volumes at a constant level then a dual duct system must be used.

Single zone system. A single air handling unit will, at a given time, supply air at one temperature only. If the building has an even demand for heating or cooling it can be treated as a single zone. Because of this the conditioned air will be delivered evenly throughout the space. If the cooling load should change, for example due to computer equipment being turned on, then the need for additional cooling will be dealt with by reducing the temperature of the air leaving the air handling unit.

If other large zones within the same building have different demands for heating and cooling then they must be treated individually. This is achieved by using additional air handling units and ducting. Each one supplying air at the temperature necessary to satisfy the zone it supplies.

Variations in demand on a smaller scale requiring room

by room control must be dealt with by changing the way the conditioned air enters the space. The methods are described below.

Variable air volume. This system is used in buildings which require cooling throughout but where individual spaces need different amounts of cooling. The system, shown in figure 7.24 and using symbols shown in figure 7.23, achieves room by room control of temperatures by varying the amount of chilled air allowed to enter the room. If the room is too warm more chilled air will be allowed to enter, if the room becomes too cool the amount of chilled air entering the room will be reduced.

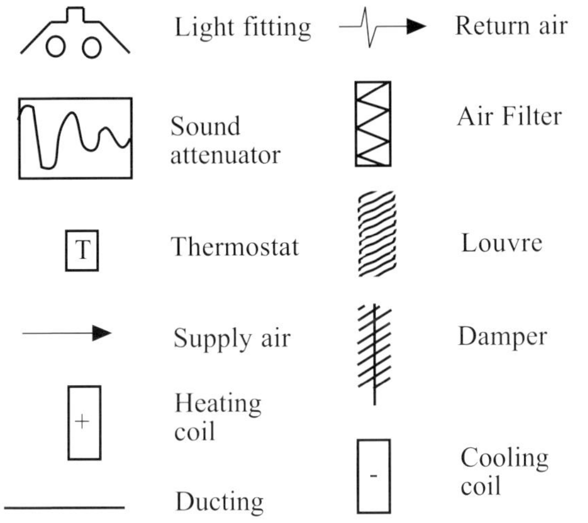

Figure 7.23 Symbols used in schematics

The central air handling unit supplies air sufficiently chilled to satisfy the maximum cooling load of the building. This is delivered to the rooms through units called variable air volume (VAV) terminals. VAV terminals (figure 7.25) control the amount of chilled air entering the room using a motorised damper. The position of this damper is determined in response to temperatures measured by a room thermostat. Depending on the room temperature the VAV terminal can vary airflow rates between zero and full flow.

An example of the control strategy begins with chilled air entering the room. This will mix with the existing room air and cause the room temperature to drop. The fall in room temperature will be detected by the room thermostat. This information is noted by the BEMS or a dedicated VAV controller which in turn will send a signal to the actuators controlling the damper position in the VAV terminal. The damper will close reducing the volume of chilled air being allowed into the room.

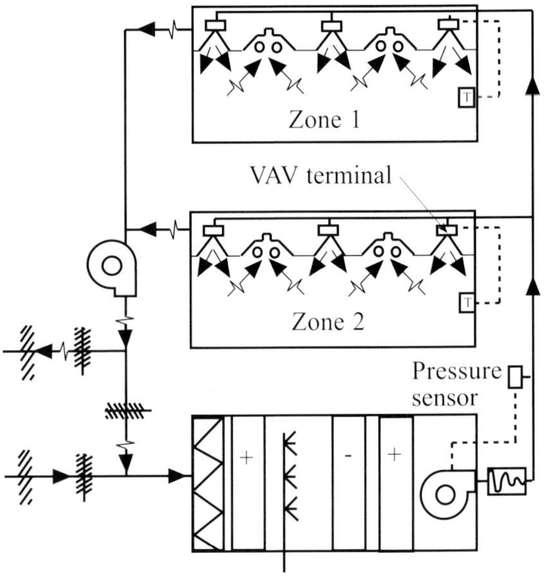

Figure 7.24 VAV air conditioning system

As the VAV dampers close the airflow from the air handling unit will be restricted and so the pressure in the ducting will rise. This is sensed using a supply duct pressure sensor which reduces the speed of the supply fan to maintain a constant pressure. In this way the fan matches the supply of air to demand.

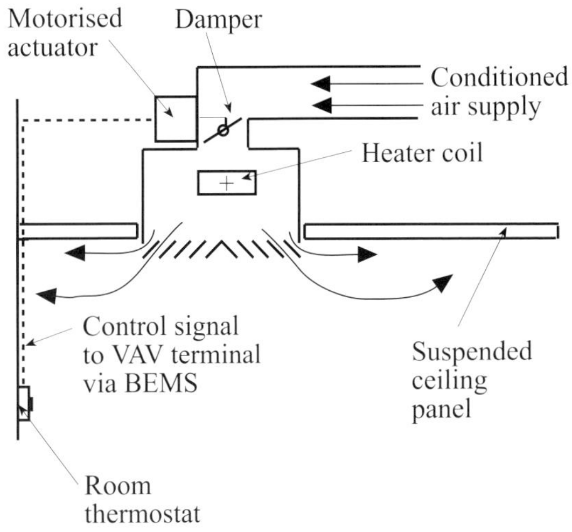

Figure 7.25 Variable air volume (VAV) terminal

If there is a need for some heating of rooms or spaces alongside the cooling system then VAV units fitted with electric heaters are used to heat up the chilled air as it leaves the unit. One application of these devices might be at the perimeter of large open plan offices where heat losses from perimeter glazing could create a local demand for heating.

Dual duct system. From the description of the VAV system above it can be seen that the volume of air entering the room varies with the demand for cooling. In some cases it is necessary to maintain a constant ventilation rate but retain close control of temperatures. This can be achieved using a constant volume system also known as a dual duct system.

The dual duct system requires two air handling units or a single air handling unit which is able to produce both chilled air and heated air at the same time (figure 7.26). Two sets of ducting are required to deliver both of these airstreams to mixing units in the rooms (figure 7.27).

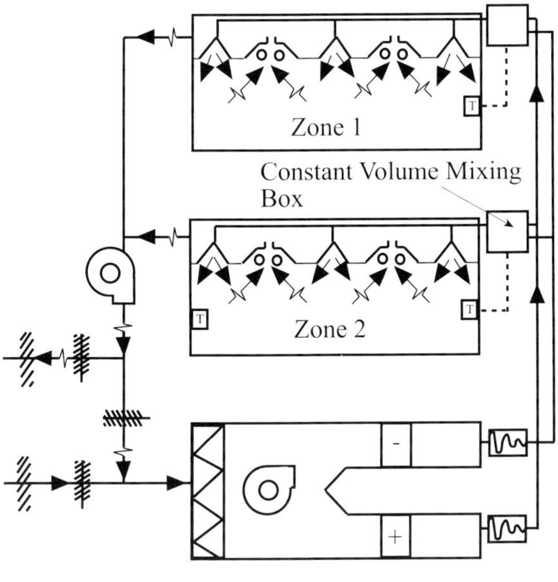

Figure 7.26 Dual duct air conditioning system

Room temperatures are controlled by varying the temperature of air entering the space by mixing the hot and cold airstreams. If cooling is required the motorised damper will allow more chilled air into the room. If heating is required the mixing box will respond to the room temperature sensor and control signals by allowing more heated air into the space. It can be seen that the system ensures that a constant volume of air always enters the space although the proportions of hot and cold air that make up this constant volume may vary. Controlling temperatures by mixing hot and cold airstreams is not a very energy efficient technique. The system does however give good delivery of ventilation and close control of temperatures.

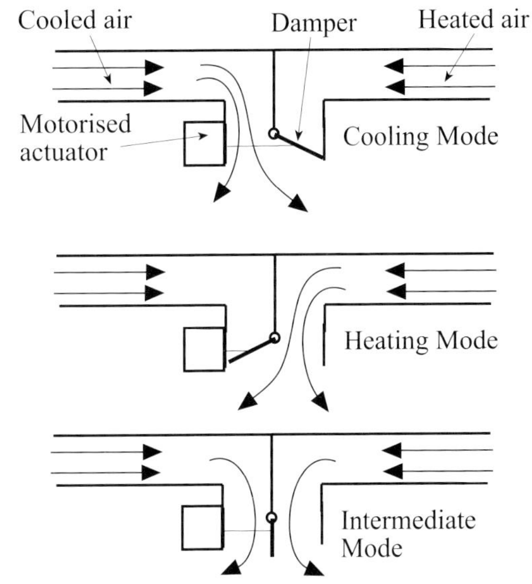

Figure 7.27 Dual duct mixing box

8.0 Partially Centralised Air/Water Systems

One of the disadvantages of centralised air conditioning systems is that the ducts needed to deliver heating or cooling to the spaces are much larger than if ventilation only were supplied to the space. This makes them difficult to accommodate within the structure especially if it is upgrading or retrofit work in an existing building. Partially centralised systems (figure 8.1) use reduced duct sizes because they only deliver enough filtered and tempered air to the rooms to satisfy the ventilation requirements. The heating or cooling demand is satisfied using room based devices. There are two types of room units, fan coil units and induction units both of which are supplied with heated water and/or chilled water from boilers and

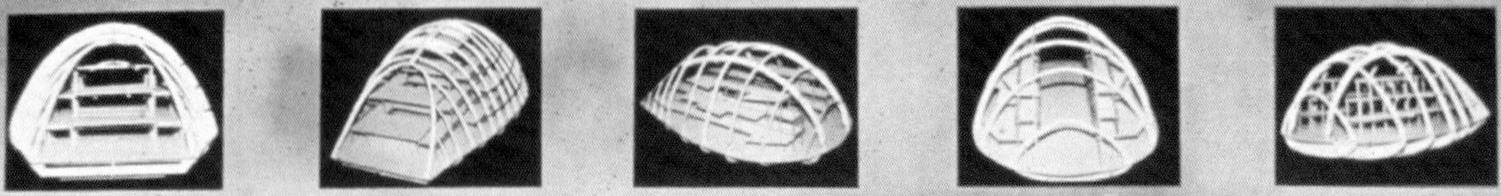

Buildings during their construction and subsequent operation consume vast amounts of natural resources. They account for half of the UK's primary energy consumption. They demand quarrying and exploitation of forests and other natural resources to supply the materials from which they are made. In use building emissions add to global warming, damage the ozone layer and create waste disposal problems.

MSc. Sustainable Architecture

One year full time - two years part time

The internal environment has been linked with ill health ranging from chronic illness caused by discomfort to life threatening illness due to the collection and concentration of pathogens and carcinogens. This course considers the tools available to alleviate these environmental and health problems such as; design methods, technologies (alternative and high tech) and legslation.

The subjects covered include: low energy design, research, management and academic methods, the global environment, health and contemporary sustainable architecture. The major component is a final research project or design thesis.

The course is aimed primarily at graduates in a discipline associated with the built environment (architecture, civil engineering, planning, building services etc.) but other disciplines will be considered.

This course offers graduates and practitioners the opportunity to expand their skills in an area of great concern both to clients and the public as a whole.

For further details contact:

Richard Nicholls
Tel 01484 472652, email R.Nicholls@hud.ac.uk
Department of Architecture, Huddersfield University
Queensgate, Huddersfield, HD1 3DH

University of
HUDDERSFIELD

chillers situated in the plant room.

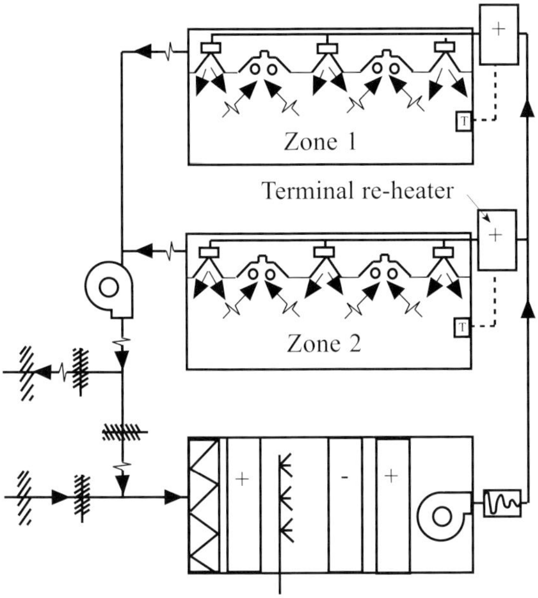

Figure 8.1 Partially centralised air/water system

Terminal units. Figure 8.2 shows a room based four pipe induction unit.

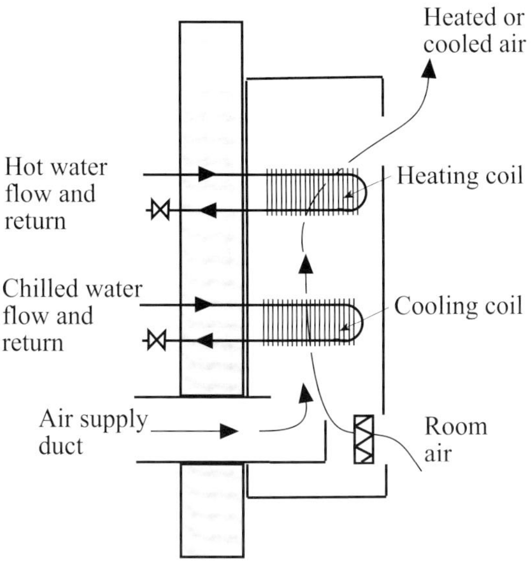

Figure 8.2 Four pipe terminal unit (induction)

Tempered ventilation air enters the unit at a relatively high speed. This incoming air drags or induces room air to also enter the unit. The mixed supply and room air then passes over a hot coil or a cold coil depending on whether there is a demand for heating or cooling. The conditioned air enters the room through an upper grille. Some units are two pipe units having a single coil. Heated or chilled water is sent through this coil as required.

Fan coil units are fitted with heating and cooling coils like the induction units but the air movement through them and mixing of supply and room air is generated using a fan and not the momentum of the air.

Displacement ventilation is a partially centralised air conditioning system which is increasingly being used in the UK (figure 8.3). In this system air is input to the room at very low velocity using raised floor terminals or low level wall terminals. The incoming air is at 18°C which is a relatively high temperature when compared to all air systems. The low airspeed and high temperature are necessary to avoid discomfort since the air is input directly into the occupied zone.

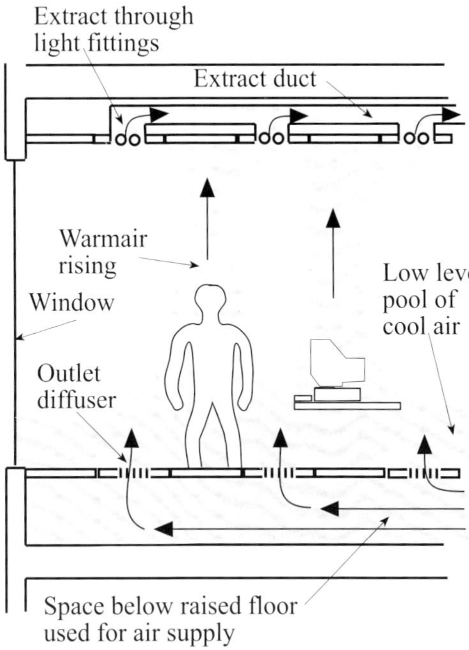

Figure 8.3 Displacement ventilation and chilled ceiling system

The supply air being cooler than the existing room air, pools in a layer along the floor. The presence of any sources of heat such as occupants bodies, office electronic equipment or pools of sunlight on the office floor will heat this pool of air causing an upward convection current to develop at the site of the source of heat. As a result fresh cool air is automatically brought to the heat source. Heat sources usually coincide with a source of pollution also. Occupants for example give out heat, metabolic CO_2 and odours. The rising warm, stale air from these sources is extracted at high level.

Because of the relatively high input air temperature displacement ventilation cannot satisfy very high cooling loads and so it is often used in conjunction with a chilled ceiling or chilled beam cooling system.

Chilled ceilings (figure 8.4) are composed of an array of purpose built suspended ceiling panels. The panels are of a standard size and made out of perforated aluminium sheet. A coil of copper pipe is fixed, in close contact, to the back of this panel. When chilled water is circulated through this pipe the ceiling panel becomes chilled. As a result any air in contact with the ceiling will become cooled and descend into the room. The room occupants will also feel cooler because their bodies will radiate heat to the chilled ceiling making them feel cool. This is the opposite effect to being stood next to a hot radiator.

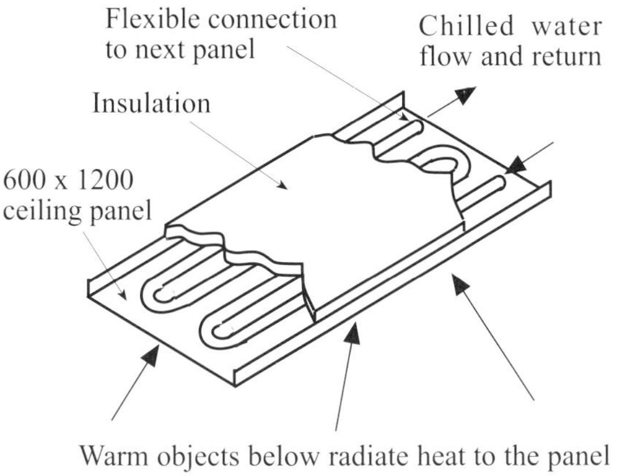

Figure 8.4 Chilled ceiling panel

There is a risk that condensation will form on the chilled ceiling. To avoid this the chilled ceiling control system must monitor humidity levels within the space. If the humidity levels indicate a risk of condensation occurring then either the incoming air must be dehumidified or the chilled ceiling surface temperature must be raised.

Chilled beams can also be used in conjunction with displacement ventilation. A passive chilled beam is shown in figure 8.5. It can be seen that the chilled surface is formed into a linear finned coil, this coil is then surrounded by a pressed steel casing and is suspended from the ceiling. Warm room air rises to the ceiling and enters the top of the beam . It is then cooled by contact with the cold coil. The cool air descends into the room through outlet slots on the underside of the beam. It can be seen that chilled beams cool a room entirely by convection.

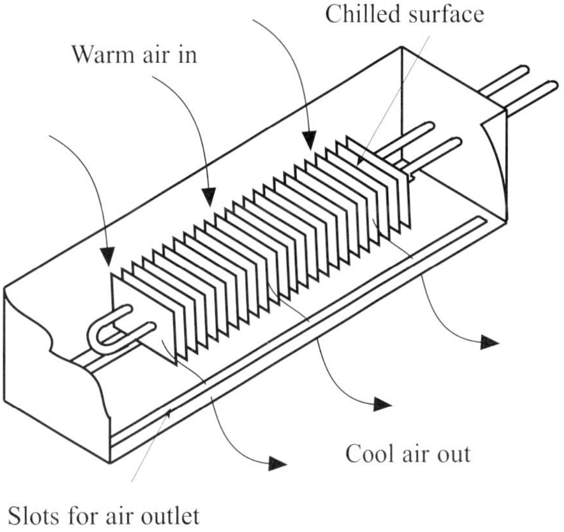

Figure 8.5 Passive chilled beam

As the cooling output of a chilled beam increases, say by reducing the water flow temperature through the device, there is a possibility that the beam will create uncomfortable cold down draughts. One way of overcoming this problem is to use active chilled beams. Active chilled beams are not used in conjunction with displacement ventilation, instead the tempered ventilation air is supplied through ducting within the beam itself. This is illustrated in figure 8.6 which shows a section through an active chilled beam. Tempered air leaves the supply ducting through slots or nozzles with sufficient velocity that it drags (induces) warm room

Mechanical & Electrical Software

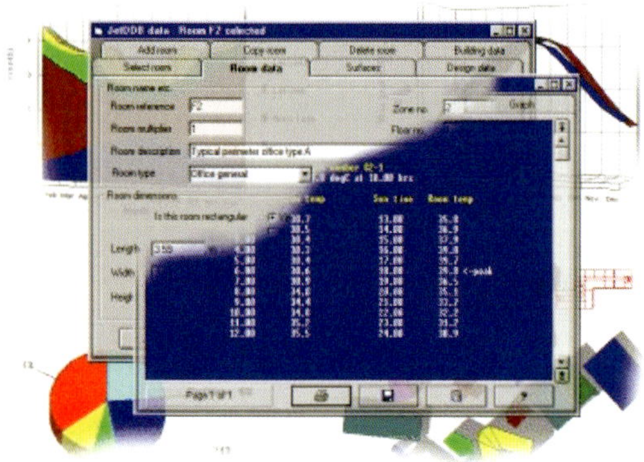

Hevacomp provides a comprehensive package of building services design software for Mechanical and Electrical services and CAD. With over 2300 sites using our design software, Hevacomp is the industry standard package. With over 18 years experience in the building services software industry, Hevacomp is the engineer s choice.

Design packages include:

- Mechanical design for load calculations, pipe and duct sizing and mechanical CAD.
- Electrical design conforming to the requirements of IEE 16th Edition wiring regulations, lighting systems design and electrical CAD.
- Support and training focused on satisfying the increased technical demands on M&E design engineers.
- Mechanical and Electrical, 3D and 2D, CAD libraries for drafting.

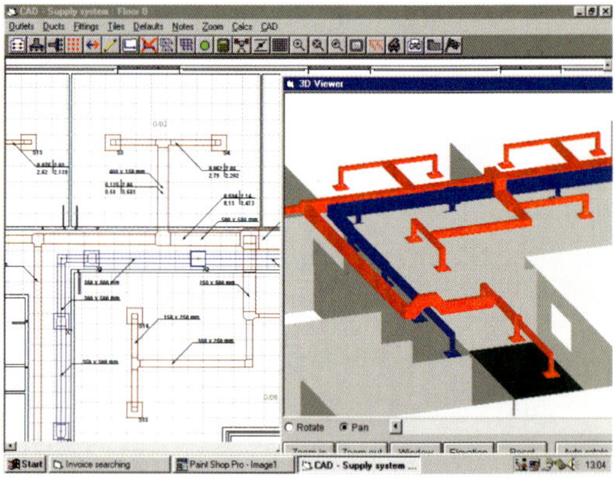

Load calculations such as heat loss, radiator sizing, heat gains (CIBSE, Carrier and ASHRAE), shadow analysis, heating and air conditioning energy, summer overheating, Room data can be set up automatically from an imported DXF CAD drawing or directly entered. Room selection for load calculations is either graphical from the drawing or by direct selection.

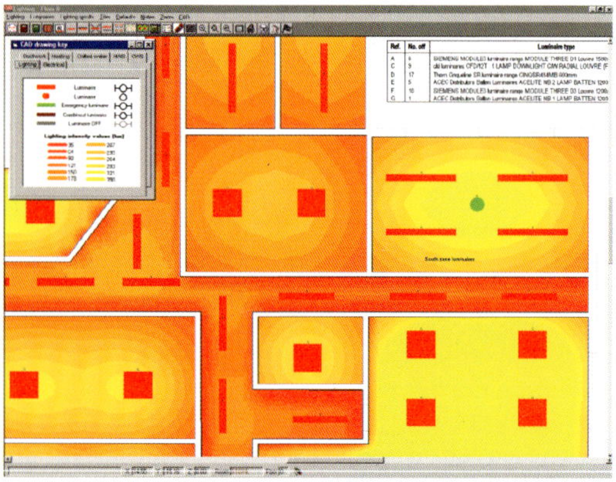

Systems can be sketched out schematically on screen, forming the data input to Hevacomp s pipe and duct sizing programs. This powerful graphical interface provides a simple and intuitive input method which enables systems to be defined faster and with less error. Systems can be sketched in plan or as an isometric.

Hevacomp s mechanical CAD package enables heating, cooling, piped services, electrical and lighting systems to be designed directly on to CAD drawings produced by AutoCAD or any other CAD system. A project can extend over multiple floors and systems can be drawn and sized across floors. Duct and pipework schematics, isometrics, full 2D drawings and 3D visualisations are reproduced.

HEVACOMP Ltd., Smitheywood House
Smitheywood Crescent, Sheffield, S8 0NU
Tel. 01142 556680
Fax. 01142 556638
www.hevacomp.com

HEVACOMP

air into the beam and through the cooling coil reducing its temperature. The supply and chilled room air mix and enter the room via outlet slots on the underside of the beam. The velocity with which the air leaves the inclined slots is sufficiently high to project it horizontally into the room above the occupied space. In this way cooler airstreams can be used without creating a cold draught in the occupied zone.

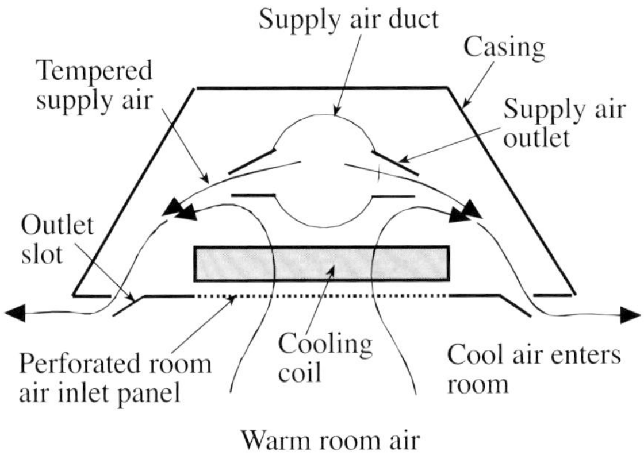

Figure 8.6 Active chilled beam

Chilled ceilings and beams are a low maintenance method of cooling a room. There are no internal fans or filters that could break down or need cleaning. The fin spacing on chilled beams means that dust build up can be largely ignored.

Energy issues. One of the benefits of systems incorporating chilled ceilings or chilled beams with displacement ventilation or active chilled beams alone is that they are an energy efficient method of cooling. This arises due to the operating parameters of the system. The first is the low fan speed used to deliver air to the outlet diffusers. Information panel IP11 explains how reducing the fan speed gives considerable reductions in fan motor energy consumption. Secondly, the chilled ceiling and beams operate at a relatively high chilled water flow temperature. This means that the chiller has to do less work and therefore will consume less electricity. The coefficient of performance of the chiller is also improved by approximately 20% due to the higher evaporator temperature(see information panel IP20, page 94).

In addition, the high operating temperatures allow the use of free and natural cooling. On the air side, if the outside air is sufficiently cool, it can be brought into the building as the supply. Since it does not have to be tempered by chillers it is known as free cooling. On the water side, it is possible to achieve the flow temperatures required using evaporative cooling. This is when water is chilled by natural evaporation in a cooling tower (page 97) and is used to supply the chilled beam or ceiling.

Q U E S T I O N S

INDIRECT.HEATING

1. **Which of the following is not a function of pipe insulation?**
- ☐ Preventing heat loss from hot pipes
- ☐ Protecting the pipe from impacts
- ☐ Preventing heat gain by cold pipes
- ☐ Preventing condensation on cold pipes

2. **Which of the following is not related to indirect heating**
- ☐ Single flue
- ☐ Heats one room only
- ☐ Single fuel supply
- ☐ Centralised control

3. **Which of the following is not related to direct heating?**
- ☐ Stand alone heaters.
- ☐ Complex control of many rooms.
- ☐ Individual fuel supply to each heat emitter.
- ☐ Heat distribution using water as a medium

4. **Which of the following categories of heating system considerations involves avoiding the release of asphyxiant gasses?**
- ☐ Economics
- ☐ Comfort
- ☐ Environment
- ☐ Safety

5. **What would be the boiler power needed for a typical detached house?**
- ☐ 5 kW
- ☐ 10 kW
- ☐ 15 kW
- ☐ 20 kW

6. **Which of the following devices prevents unburnt gas building up in a boiler?**
- ☐ Flame failure device
- ☐ Pilot light
- ☐ Boiler thermostat
- ☐ Heat exchanger

7. **Which of the following is not an effect of acid rain?**
- ☐ Atmospheric warming
- ☐ Leaf damage
- ☐ Damage of freshwater life
- ☐ Erosion of statues

8. **Which of the following prevents a reversal of flue gasses through the boiler on windy days?**
- ☐ Fan dilution
- ☐ Flue terminal
- ☐ Ventilation openings
- ☐ Draught diverter

9. **When the load on a gas boiler decreases the efficiency**
- ☐ Increases
- ☐ Stays the same
- ☐ Decreases
- ☐ Becomes unstable

10. **Which of the following does not help maximise CHP running hours?**
- ☐ Sizing of heat output to match base loads
- ☐ Export electricity meters
- ☐ Installing the CHP as lead heat source
- ☐ Routine maintenance

11. **What is the efficiency of a typical 3kW pump motor?**
- ☐ 61%
- ☐ 71%
- ☐ 81%
- ☐ 91%

12. **Which of the following radiators would give the highest heat ouput for a given area?**
- ☐ Double panel
- ☐ Double convector
- ☐ Single panel
- ☐ Single convector

13. When a person is near, but not touching, a cold window they experience

- ☐ Radiant heat losses
- ☐ Convective heat gains
- ☐ Conductive heat losses
- ☐ Evaporative heat losses

14. Which of the following is usually only encountered in commercial buildings?

- ☐ Combi boilers
- ☐ Water to water plate heat exchangers
- ☐ Indirect cylinders
- ☐ Gas fired water heaters

15. When a substance absorbs sensible heat it

- ☐ Changes from a solid to a liquid
- ☐ Increases in temperature
- ☐ Changes from a liquid to a gas
- ☐ Decreases in temperature

16. What percentage of total floor area is taken up by services in a speculative air-conditioned office?

- ☐ 4 - 5%
- ☐ 6 - 9%
- ☐ 10 - 15%
- ☐ 15 - 30%

17. Which of the following gives room by room control of temperatures?

- ☐ Boiler thermostat
- ☐ Thermostatic radiator valves
- ☐ Zone thermostat
- ☐ dhw cylinder thermostat

18. Decreasing boiler flow temperatures as outside air temperatures increase is known as

- ☐ Compensation
- ☐ Boiler step control
- ☐ Optimisation
- ☐ Boiler cycling

19. Which of the following may not require its heating system to be zoned?

- ☐ A large cellular office building with south facing glazing
- ☐ A small open plan office
- ☐ A school holding night classes
- ☐ An office building with a computer suite

20. Which of the following devices opens and closes valves?

- ☐ Outstations
- ☐ Actuators
- ☐ Sensors
- ☐ Supervisor

21. Which of the following is not a function of a BEMS

- ☐ Optimum start/stop timing
- ☐ Adjustment of set points
- ☐ Fault reporting
- ☐ Making the tea

22. Which of the following valves is used to isolate faulty components?

- ☐ Globe valve
- ☐ Three port valve
- ☐ Two port valve
- ☐ Butterfly valve

23. What is the typical temperature range of MTHW?

- ☐ 35 - 70°C
- ☐ 70 - 100°C
- ☐ 100 - 120°C
- ☐ 120 - 150°C

24. For the same amount of heat transfer air ducts are

- ☐ Smaller than steam pipes
- ☐ Smaller than hot water pipes
- ☐ The same size as steam and hot water pipes
- ☐ Bigger than steam and hot water pipes

DIRECT. HEATING

1. **Which of the following heat transfer methods does not require a transfer medium?**
- ☐ Conduction
- ☐ Mass transfer
- ☐ Convection
- ☐ Radiation

2. **How is the heat output of a fan assisted electric storage heater regulated?**
- ☐ By switching the fan on and off as required
- ☐ By having no insulation in the casing
- ☐ By turning the heating current on and off
- ☐ By closing the damper whilst the fan is running

3. **Commercial warm air cabinet heaters should not be used when**
- ☐ The space to be heated is draughty
- ☐ The space to be heated is not draughty
- ☐ People are working in the space
- ☐ Destratification fans are installed

4. **Which of the following does not apply to a roof mounted heating and ventilation direct heater**
- ☐ Can recirculate air in the space
- ☐ Can provide free cooling in summer
- ☐ Can assist in destratification
- ☐ Cannot supply fresh air

5. **Which of the following would be an inappropriate use for a high temperature radiant heater?**
- ☐ In rooms with low ceilings (<3m)
- ☐ In rooms with high ceilings (>3.5m)
- ☐ In draughty rooms
- ☐ Where spot heating is required

6. **What is the approximate operating efficiency of a direct fired water heater?**
- ☐ 70%
- ☐ 80%
- ☐ 90%
- ☐ 100%

VENTILATION

1. **Which of the following is not a function of ventilation?**
- ☐ To raise the relative humidity
- ☐ To supply oxygen for breathing
- ☐ To dilute pollutants
- ☐ To remove unwanted heat

2. **The dilution of body odours from sedentiary occupants requires a ventilation rate of**
- ☐ 40 l/s
- ☐ 32 l/s
- ☐ 16 l/s
- ☐ 8 l/s

3. **Infiltration is**
- ☐ Ventilation using ducts
- ☐ Uncontrolled natural ventilation
- ☐ Fan driven ventilation
- ☐ Ventilation through a vertical tube

4. **A humidistat turns off the bathroom extract fan when**
- ☐ The light is switched off
- ☐ The room air moisture content is satisfactory
- ☐ The room goes cold
- ☐ Condensation forms on the windows

5. **Which of the following is least likely to need mechanical ventilation?**
- ☐ A room where the volume per person is <$3.5m^3$
- ☐ Deep plan rooms away from outside walls
- ☐ Shallow rooms with outside wall and windows
- ☐ Shallow room with sealed winows

6. **Which of the following involves the fanned supply and extract of air to rooms?**
- ☐ Balanced Ventilation
- ☐ Supply ventilation
- ☐ Extract ventilation
- ☐ Passive stack ventilation

7. **Which of the following fan types changes the direction of airflow by 90°?**
- ☐ Axial flow fan
- ☐ Propeller fan
- ☐ Centrifugal fan
- ☐ Extract fan

8. Which of the following air to air heat recovery methods has no moving parts?
- ☐ Run around coils
- ☐ Thermal wheel
- ☐ Plate heat exchanger
- ☐ Heat pump

AIR-CONDITIONING

1. Which of the following rooms is least likely to need air conditioning?
- ☐ Office with large south facing windows
- ☐ Computer suite
- ☐ An operating theatre
- ☐ Domestic living room

2. Which of the components of a vapour compression chiller is used to reject heat extracted during cooling?
- ☐ Evaporator coil
- ☐ Compressor
- ☐ Expansion valve
- ☐ Condenser coil

3. Which of the following refrigerants is most destructive to the ozone layer if it escapes?
- ☐ R11 - CFC
- ☐ R22 - HCFC
- ☐ R134a - HFC
- ☐ Ammonia

4. What is the coefficient of performance (COP) of a typical heat pump?
- ☐ 2.0
- ☐ 1.0
- ☐ 4.0
- ☐ 3.0

5. A split air conditioning unit is so called because
- ☐ The refrigerant is divided by the expansion valve
- ☐ Chilled air output is bi-directional
- ☐ It has a seperate indoor and an outdoor unit
- ☐ It sits on the window sill, part inside and part outside

6. The indoor and outdoor units in a variable refrigerant flow system can be seperated by
- ☐ 100m including a vertical rise of 50m
- ☐ 100m including a vertical rise of 10m
- ☐ 50m including a vertical rise of 10m
- ☐ 50m with no vertical rise

7. At the heart of a centralised air-conditioning system is the
- ☐ Diffuser
- ☐ Filter
- ☐ Air handling unit
- ☐ Damper

8. Which of the following filters has the largest dust carrying capacity?
- ☐ Panel filters
- ☐ HEPA filters
- ☐ Bag filters
- ☐ Biological filter

9. The only type of filter which can remove gasses and odours is a
- ☐ Roll filter
- ☐ Electrostatic filter
- ☐ Pre-filter
- ☐ Activated carbon filter

10. An absorption chiller
- ☐ Has a COP higher than a vapour compression chiller
- ☐ Uses a gas burner or waste heat to drive it
- ☐ Uses HCFC's as a refrigerant
- ☐ Use electricity to drive a compressor

11. Which of the following does not apply to air cooled condensers
- ☐ Must have a stream of air flowing across it
- ☐ Requires a constant spray of water
- ☐ Is less efficient than an evaporative condenser
- ☐ Has the advantage of not requiring water

12. Legionnaires disease is associated with poorly maintained
- ☐ Window sill air conditioners
- ☐ Air cooled condensers
- ☐ Split air conditioning units
- ☐ Wet heat rejection equipment

13. For human comfort the relative humidity in a room should be in the range
- ☐ 10 - 20% RH
- ☐ 40 - 70% RH
- ☐ 20 - 40 % RH
- ☐ 70 - 90% RH

14. The most energy efficient wet humidification system is a
- ☐ Capillary washer
- ☐ Atomizing nozzle humidifier
- ☐ Air washers
- ☐ Ultrasonic humidifier

15. Steam humidifiers operate adiabatically. This means
- ☐ The air temperature decreases
- ☐ The air temperature stays the same
- ☐ The air pressure increases
- ☐ The air pressure decreases

16. Dehumidification by chilling is energy intensive. This is because
- ☐ The air must be cooled to 0°C
- ☐ The air must be heated to 100°C
- ☐ The air must be heated to dry it then re-cooled
- ☐ The air is cooled below dew point then re-heated

17. Which of the following statements regarding the entry of conditioned air into a room using diffusers is incorrect?
- ☐ The air should enter quietly
- ☐ The air should mix outside of the occupied zone
- ☐ The air should enter the occupied zone immediately
- ☐ The air should ventilate all parts of the room

18. The mechanism which increases the distance travelled by air leaving a suspended ceiling diffuser is called?
- ☐ The entrainment effect
- ☐ The throw effect
- ☐ The swirl effect
- ☐ The coanda effect

19. Which of the following locations is unsuitable for a room air extract grille
- ☐ Within a stagnant air zone
- ☐ Next to the supply diffuser
- ☐ Near a known source of pollution
- ☐ Where light dust staining of surfaces is acceptable

20. Which of the following does not increase duct airflow resistance
- ☐ Keeping ducting linear
- ☐ Bends
- ☐ Dampers
- ☐ Reducing the cross sectional area

21. Which type of damper is used to change the direction of airflow
- ☐ Butterfly damper
- ☐ Opposed blade damper
- ☐ Fire damper
- ☐ Parallel blade damper

22. Which centralised air conditioning delivery system offers room by room control of temperatures whilst maintaining ventilation
- ☐ Variable air volume (VAV) system
- ☐ Dual duct system
- ☐ Variable air volume incorporating electric heating
- ☐ Single zone system

23. In a displacement ventilation system
- ☐ Cool air is input from ceiling diffusers
- ☐ Cool air is input at low level
- ☐ Warm air is input at low level
- ☐ Warm air is input from ceiling diffusers

24. A chilled ceiling cools by
- ☐ Convection and thermal radiation
- ☐ Convection only
- ☐ Thermal radiation only
- ☐ Conduction

25. Active chilled beams differ from passive chilled beams in that
- ☐ They have inbuilt fans
- ☐ They supply ventilation alongside comfort cooling
- ☐ They cool by convection and thermal radiation
- ☐ They have a lower cooling capacity

ENERGY.EFFICIENCY

The following questions relate to the information given on the keeping tabs on energy efficiency advice pages

1. Which of the following is not a recommendation for reducing lighting energy consumption?

- ☐ Lighting to appropriate levels but not more
- ☐ Use efficient electric lighting and provide good controls
- ☐ Leave lights on in rooms at all times
- ☐ Make good use of daylight

2. Which of the following is not a method for reducing space heating energy consumption?

- ☐ Use night time ventilation
- ☐ Provide adequate building insulation
- ☐ Specify efficient space heating equipment
- ☐ Provide effective controls

3. Which of the following forms of energy produces the most carbon dioxide per kWh

- ☐ Solar energy
- ☐ Oil
- ☐ Gas
- ☐ Electricity

4. Energy use in buildings accounts for which of the following percentages of UK carbon dioxide output?

- ☐ 4%
- ☐ 40%
- ☐ 45%
- ☐ 54%

5. Which of the following building types is most likely to benefit from a CHP system?

- ☐ Factories
- ☐ Leisure centres with pools
- ☐ Houses
- ☐ Cinemas

6. Which of the following is not a consideration in holistic low energy design?

- ☐ Low cost forms of energy
- ☐ The building form
- ☐ The building services
- ☐ Post occupancy monitoring and targetting

7. When pressure tested for air leakage the target range for leakage in homes is

- ☐ less than 7.5 m^3/h
- ☐ 7.5 to 15 m^3/h
- ☐ 15 to 20.5 m^3/h
- ☐ 20.5 to 25 m^3/h

8. Which of the following is not a benefit of increased airtightness of buildings?

- ☐ Less discomfort due to draughts
- ☐ Reduced energy costs
- ☐ Reduced pump size
- ☐ Reduced heating plant size

9. Using larger ducts with slower air speeds reduces the energy consumption of mechanical ventilation systems by?

- ☐ 70%
- ☐ 50%
- ☐ 30%
- ☐ 10%

10. Which of the following is not an element in a passive cooling strategy?

- ☐ Reduce solar heat gains
- ☐ Use a vapour compression chiller
- ☐ Reduce equipment and lighting heat gains
- ☐ Leave structural mass exposed to the room air

INDEX

A

B

C

D

I

L

M

O

P

R

S

DIRECTORY.OF INDUSTRIAL.SPONSORS

Belimo Automation (UK) Ltd. page 42, 112
The Lion Centre
Hampton Road West
Feltham
Middlesex
TW13 6DS

Tel. 0181 755 4411
Fax. 0181 755 4042
www.belimo.org
email: belimo@belimo.co.uk

Building Controls Group page 40
P.O. Box 1397
Highworth
Swindon
SN6 7UD

Tel. 01793 763556
Fax. 01793 763556
www.esta.org.uk/bcg
email: bcg@esta

Action Energy page 2, 8, 16, 18
62, 64, 70

Tel. 0800 585794
www.actionenergy.org.uk

Calorex Heat Pumps Ltd. page 104
The Causeway
Maldon
Essex
CM9 5PU

Tel. 01621 856611
Fax. 01621 850871
www.calorex.com
email: sales@calorex.com

Gilberts (Blackpool) Ltd page 106
Head Office and Works
Clifton Road
Blackpool
Lancashire
FY4 4QT

Tel. 01253 766911
Fax. 01253 767941
www.gilbertsblackpool.com
email: sales@gilbertsblackpool.com

Hamworthy Heating Ltd. page 14
Fleets Corner
Poole
Dorset
BH17 0HH

Tel. 01202 662552
Fax. 01202 665111
email. sales@hamworthy-heating.com
www.hamworthy-heating.com

JS Humidifiers plc. page 98
Rustington Trading Estate
Artex Avenue
Rustington
Littlehampton
West Sussex
BN16 3LN

Tel. 01903 850 200
Fax. 01903 850 345
www.jshumidifiers.com
email: specguide@humidifiers.co.uk

Johnson Control Systems Ltd. page 36
Johnson controls House
Randalls Research Park
Randalls Way
Leatherhead
Surrey
KT22 7TS

Tel. 01372 370400
Fax. 01372 376823
www.johnsoncontrols.com
email: neil.stuart@jci.com

Siemens Building Technology Ltd. page 34
Landis and Staefa Division
Hawthorne Road
Staines
Middlesex
TW18 3AY

Tel. 01784 461616
Fax. 01784 464646
www.landisstaefa.co.uk

Trend Controls Ltd. page 38
PO Box 34
Horsham
West Sussex
RH12 2YF

Tel. 01403 211888
Fax. 01403 241608
www.trend-controls.com
email: careersinfo@caradon.com

University of Huddersfield page 120
Department of Architecture
Queensgate
Huddersfield
HD1 3DH

Tel. 01484 472289
Fax. 01484 472440
www.hud.ac.uk/pg_pros/index.html
email: sustainable.architecture@hud.ac.uk

University of Northumbria page 118
Engineering, Science and
Technology Faculty Office
Ellison Building
Ellison Place
Newcastle upon Tyne
NE1 8ST

Tel. 0191 227 4453
Fax. 0191 227 4561
http://online.northumbria.ac.uk/faculties/est/
built_environment
email:et.admissions@northumbria.ac.uk

Manchester School of Architecture page 116
Manchester Metropolitan University
Chatham Building
Cavendish Street
Manchester
M15 6BR

Tel. 0161 247 6959
email: g.keeffe@mmu.ac.uk

South Bank University page 122
Division of Energy, Environment and Building Services
103 London Road
London
SE1 0AA

Tel. 020 7815 7612
www.sbu.ac.uk/sesd
email: afekare@sbu.ac.uk

Hevacomp Ltd. page 124
Smitheywood House
Smitheywood Crescent
Sheffield
S8 0NU

Tel. 01142 556680
Fax. 01142 556638
www.hevacomp.com
email: sales@hevacomp.com